Sushil C. Dimri, Preeti Malik, Mangey Ram
Algorithms

W0234616

Also of interest

Data Structures and Algorithms Analysis – New Perspectives.
Vol. 1 Data structures based on linear relations
Xingni Zhou, Zhiyuan Ren, Yanzhuo Ma, Kai Fan and Xiang Ji, 2020
ISBN 978-3-11-059557-4, e-ISBN (PDF) 978-3-11-059558-1,
e-ISBN (EPUB) 978-3-11-059318-1

Data Structures and Algorithms Analysis – New Perspectives.
Vol. 2 Data structures based on non-linear relations and data
processing methods
Xingni Zhou, Zhiyuan Ren, Yanzhuo Ma, Kai Fan and Xiang Ji, 2020
ISBN 978-3-11-067605-1, e-ISBN (PDF) 978-3-11-067607-5,
e-ISBN (EPUB) 978-3-11-067616-7

Multivariate Algorithms and Information-Based Complexity
Fred J. Hickernell and Peter Kritzer (Eds.), 2020
ISBN 978-3-11-063311-5, e-ISBN (PDF) 978-3-11-063546-1,
e-ISBN (EPUB) 978-3-11-063315-3

Computational Intelligence.
Theoretical Advances and Advanced Applications
Dinesh C.S. Bisht and Mangey Ram (Eds.), 2020
ISBN 978-3-11-065524-7, e-ISBN (PDF) 978-3-11-067135-3,
e-ISBN (EPUB) 978-3-11-066833-9

Machine Learning Applications.
Emerging Trends
Rik Das, Siddhartha Bhattacharyya and Sudarshan Nandy (Eds.), 2020
ISBN 978-3-11-060853-3, e-ISBN (PDF) 978-3-11-061098-7,
e-ISBN (EPUB) 978-3-11-060866-3

Sushil C. Dimri, Preeti Malik, Mangey Ram

Algorithms

———

Design and Analysis

DE GRUYTER

Authors
Sushil C. Dimri
Department of Computer Science and Engineering
Graphic Era
566/6 Bell Road
Clement Town 248002, Dehradun, Uttarakhand, India
dimri.sushil2@gmail.com

Preeti Malik
Department of Computer Science and Engineering
Graphic Era
566/6 Bell Road
Clement Town 248002, Dehradun, Uttarakhand, India
preetishivach2009@gmail.com

Mangey Ram
Department of Computer Science and Engineering
Graphic Era
566/6 Bell Road
Clement Town 248002, Dehradun, Uttarakhand, India
drmrswami@yahoo.com

ISBN 978-3-11-069341-6
e-ISBN (PDF) 978-3-11-069360-7
e-ISBN (EPUB) 978-3-11-069375-1

Library of Congress Control Number: 2020940232

Bibliographic information published by the Deutsche Nationalbibliothek
The Deutsche Nationalbibliothek lists this publication in the Deutsche Nationalbibliografie;
detailed bibliographic data are available on the Internet at http://dnb.dnb.de.

© 2021 Walter de Gruyter GmbH, Berlin/Boston
Cover image: farakos/iStock/Getty Images Plus
Typesetting: Integra Software Services Pvt. Ltd.
Printing and binding: CPI books GmbH, Leck

www.degruyter.com

Preface

Design and analysis of algorithm is a difficult subject for students as well as faculty members, especially for those who do not have sound knowledge of mathematics. Students require a simple, perfect explanation of the topic and algorithms. The book covers all mathematical aspects related to the design of algorithm. The text of the book is quite easy; every chapter contains practice question, which is sufficient to clear the concept of the chapter. The book, which we are presenting, is based on our lectures on the subject, in which we explain almost all the topics in a simple systematic manner but comprehensive.

This book contains all necessary mathematics background and clears the concepts of the students.

The organizations of chapters are as follows:

Chapter 1: Basic knowledge of mathematics, relations, recurrence relation and solution techniques, function and growth of functions.

Chapter 2: Different sorting techniques and their analysis.

Chapter 3: Greedy approach, dynamic programming, branch-and-bound techniques, backtracking and problems, amortized analysis, and order statics.

Chapter 4: Graph algorithms, breadth-first search, depth-first search, spanning tree, flow-maximization algorithms, and shortest path algorithms.

Chapter 5: Binary search tree, red black tree, binomial heap, B-tree, and Fibonacci heap.

Chapter 6: Approximation algorithms, sorting networks, matrix operations, fast Fourier transformation, number-theoretic algorithm, computational geometry randomized algorithms, string matching, NP-hard, NP-complete, and Cook's theorem.

https://doi.org/10.1515/9783110693607-202

Contents

Chapter 1
Introduction

1.1 Algorithm

An algorithm is a finite sequence of computational steps that produces the desired output when applied on a certain problem with necessary input values. In algorithm, every step must be clear, simple, definite, and unambiguous. Figure 1.1 shows diagrammatic representation of algorithm.

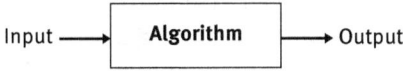

Figure 1.1: Algorithm.

1.2 Another definition

An algorithm is defined as an ultimate group of instructions to achieve a given set of goals to fulfill the given criteria.
1. **Input:** Zero or more quantities must be passed externally as input to algorithm.
2. **Output:** Minimum one quantity must be formed as an output.
3. **Finiteness:** An algorithm must be well defined in a finite set of steps.
4. **Definiteness:** Each and every instruction of a given algorithm must be precise and unambiguous.
5. **Effectiveness:** Every statement/instruction specifically contributes something in solution, defined by an algorithm.

Example 1.1: We will discuss an example of a sorting algorithm.
Input: Set of elements (sequence)
$$\{a_1, a_2 \ldots \ldots a_n\}$$
where elements are in random fashion.
Output: Elements in ascending order, that is:
$$a_i \leq a_{i+1}, \text{ for } i = 1 \text{ to } n-1.$$

1.3 Analyzing the performance of algorithms

Analysis of the algorithms allows us to take decisions about the value of one algorithm over another. When an algorithm is executed in computer, it requires central

https://doi.org/10.1515/9783110693607-001

processing unit (CPU, for performing its operations) and memory (for storing program and data). There are two parameters on which we can analyze the performance of an algorithm. These are *time complexity* and *space complexity*. As compared to space analysis, the analysis of time requirements for an algorithm is important, but whenever necessary, both the complexities are used.

The amount of memory required by a program to run to conclusion is known as *space complexity*. Space is a kind of resource that is reusable, and we can easily increase the space (memory) of the system. But time is a resource that is limited, not reusable, and nonpurchasable.

Time complexity completely relies on the size of the input. The sum of time of CPU needed by the algorithm (program) to execute to conclusion is known as *time complexity*. It should be noted that the same algorithm can take different time to run. There are three kinds of time complexity: best, worst, and average case complexity of the algorithm. *Best-case complexity* is the minimum amount of time taken by the algorithm for n-sized input. *Average-case complexity* is the time taken by the algorithm having typical input data of size "n." *Worst-case complexity* is the maximum amount of time required by the algorithm for n-sized input. These terms will be discussed later in this chapter.

1.4 Growth of the functions

The time complexity of an algorithm is generally a function of the input size "n." Growth order of a function indicates how fast a function grows with respect to change in input size "n." Some useful popular functions are:

Linear function: **n**

Square function: $\mathbf{n^2}$

Exponential function: $\mathbf{2^n}$

Logarithmic function: $\mathbf{log_2 n}$

Figure 1.2 shows how various functions grow with the input size "n." The graph is evidence for how fast exponential function grows in comparison to other functions.

1.5 Asymptotic notations

Whenever we learn about an algorithm, we are keen to depict them as per their preciseness and efficiency. In other words, we are mainly concerned about the growth of the running time of an algorithm (not in the exact running time). Scientifically, this phenomenon is called *asymptotic running time*. There is an acute need to design a method to find out the rate of growth of functions so that we can compare algorithms.

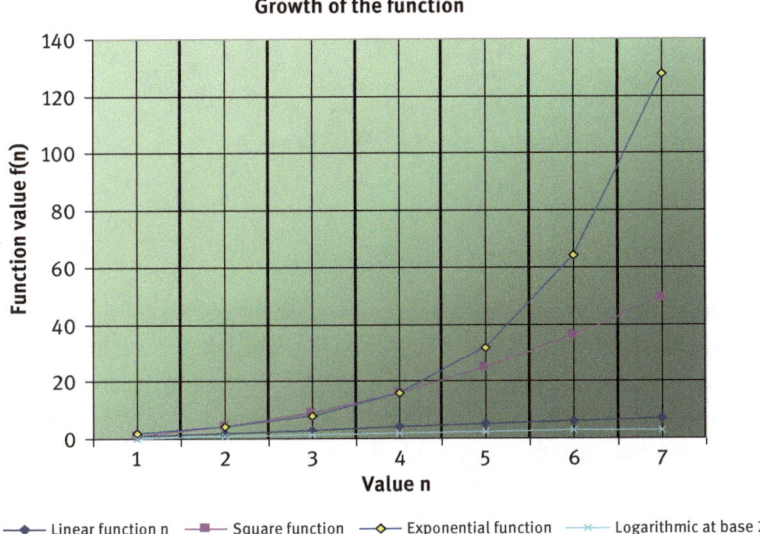

Figure 1.2: Plotting function values.

Asymptotic notation is an innovative way to provide the suitable methodology for categorizing functions according to the rate of growth of function.

1.5.1 The O (big oh) notation

Let f(n) and g(n) are two positive functions. A function f(n) is said to be in O(g(n)) signified by f(n) = O(g(n)). If f(n) is bounded above by some positive constant multiple of g(n) for large values of n, i.e. there exists some positive constant c and some nonnegative whole number n_0 (allude Figure 1.3) with the end goal that

$$f(n) \leq c\, g(n), \text{ when } n \geq n_0.$$

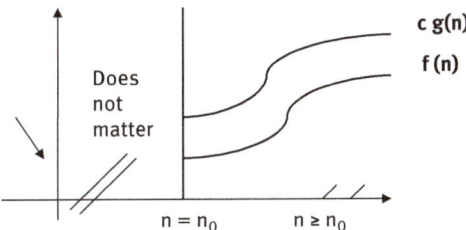

Figure 1.3: Growth of the Function f(n) and g(n) for Big oh notation.

Example 1.2:

$500\,n + 1 = f(n)$ says

$500\,n + 1 \le 500\,n + n \quad \{\forall\, n \ge 1\}$

$501\,n \le 501\,n^2$

$500\,n + 1 \le 501\,n^2,\, n \ge 1$

So $c = 501$, and $n_0 = 1$

$500\,n + 1 \in O(n^2)$

1.5.2 The Ω notation

Let f(n) and g(n) are two positive functions. The function f(n) is said to be in $\Omega(g(n))$ (Omega of g(n)) meant by $f(n) = \Omega(g(n))$); if function f(n) is lower bounded by some positive multiple of g(n) for some large value of n, i.e. there exist a positive constant c and some nonnegative whole number n_0 (allude Figure 1.4), with the end goal that

$$f(n) \ge c\,g(n), \quad \text{when } n \ge n_0$$

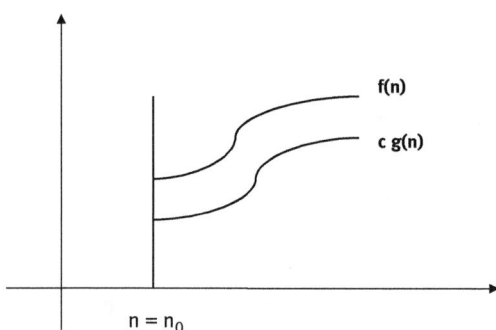

Figure 1.4: Growth of the Function f(n) and g(n) for Omega notation.

Example 1.3:

$f(n) = (n^3)$

$g(n) = (n^2)$

$n^3 \ge n^2 \,\forall\, n \ge 0$

$n^3 \ge 1.n^2, \quad \text{where } n_0 \ge 0$

Hence, $c = 1$, $n_0 = 0$, $n^3 \in \Omega(n^2)$

1.5.3 The θ notation

Let f(n) and g(n) are two positive functions. A function f(n) is said to be in $\theta(g(n))$ (Theta of g(n)), which means $f(n) = \theta(g(n))$. The function f(n) is limited above and below by some positive constant multiple of g(n) for large value of n, i.e. there exist positive constants c_1 and c_2 and some nonnegative whole number n_0 (allude Figure 1.5), with the end goal that

$$c_1\, g(n) \le f(n) \le c_2 g(n), \quad \text{when } n \ge n_0$$

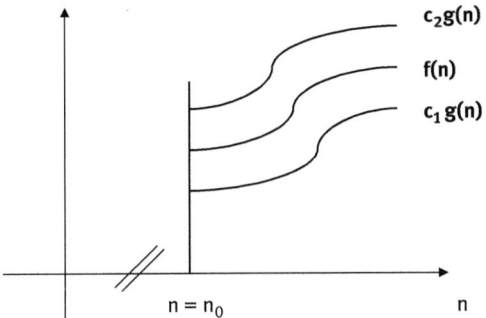

Figure 1.5: Growth of the Function f(n) and g(n) for Theta notation.

Example 1.4:

$$f(m) = \frac{1}{2}m(m-1)$$

For upper bound:

$$\frac{1}{2}m(m-1) = \frac{m^2}{2} - \frac{m}{2} \le \frac{m^2}{2}, \quad \text{where } m \ge 1$$

For lower bound:

$$\frac{1}{2}m(m-1) = \frac{m^2}{2} - \frac{m}{2} \ge \frac{m^2}{2} - \frac{m}{2}\cdot\frac{m}{2}, \quad \text{where } m \ge 2$$

$$\Rightarrow \frac{1}{2}m(m-1) \ge \frac{1}{4}m^2$$

So, we have

$$\frac{1.m^2}{4} \le \frac{1}{2}m(m-1) \le \frac{1}{2}.m^2, \quad \text{where } c_1 = \frac{1}{2}, c_2 = \frac{1}{4}$$

where $m \ge 2$

This implies that

$$\Rightarrow \frac{1}{2}m(m-1) \in \theta(m^2)$$

Theorem 1.1:

If $f_1(m) = O(g_1(m))$ and $f_2(m) = O(g_2(m))$, then $f_1(m) + f_2(m) = O(\max\{g_1(m), g_2(m)\})$

Proof:

$$f_1(m) = O(g_1(m))$$

$$\Rightarrow f_1(m) \leq c_1 g_1(m) \text{ when } m \geq m_1$$

and

$$f_2(m) = O(g_2(m))$$

$$\Rightarrow f_2(m) \leq c_2 g_2(m) \text{ when } m \geq m_2$$

taking $c_3 = \max(c_1, c_2)$

and $m_0 = \max(m_1, m_2)$

$$\Rightarrow f_1(m) + f_2(m) \leq c_1 g_1(m) + c_2 g_2(m)$$

$$\leq c_3 g_1(m) + c_3 g_2(m), \text{ when } m \geq m_0$$

$//m \geq m_0$ is common range where both inequalities are valid.

$$\leq c_3[g_1(m) + g_2(m)]\{\text{condition}(a_1, b_1), (a_2, b_2)$$

$$a_1 \leq b_1 \text{ and } a_2 \leq b_2$$

$$a_1 + a_2 \leq 2 \max(b_1, b_2)\}$$

$$\Rightarrow f_1(m) + f_2(m) \leq c_3 2\{\max(g_1(m), g_2(m)\}, \text{ when } m \geq m_0$$

$$f_1(m) + f_2(m) = O(\max\{g_1(m), g_2(m)\})$$

Theorem 1.2:

$$\text{If } t_1(m) \in O\ g_1(m) \text{ and } t_2(m) \in O\ g_2(m)$$

$$\text{Then}\{t_1(m).t_2(m)\} \in O\{(g_1(m).g_2(m)\}$$

Proof: Proof is left for you.

Use of limit for comparing growth order of function

Let $f(n)$ and $g(n)$ be two positive functions

$\lim\limits_{n \to \infty} \frac{f(n)}{g(n)} = \{$zero, means that $f(n)$ has lesser order growth than $g(n)$

$= b$, means that $f(n)$ has identical order of growth as $g(n)$

$= $ infinite, means that $f(n)$ has better order of growth than $g(n)$

Compare the order of growth of $\log_2 m$ and $\sqrt{m}$

$$\lim\limits_{m \to \infty} \{\log_2 m / \sqrt{m}\} = \lim\limits_{m \to \infty} \left\{ \frac{1}{m}\log_2 e \Big/ \frac{1}{2\sqrt{m}} \right.$$

$$= \lim_{m \to \infty} \frac{2\sqrt{m}}{m \log_2 e} = 0$$

$\Rightarrow \log_2 m$ has lesser order of growth than $\sqrt{m}$

Little o (oh) and little ω (omega) notation

Big oh (O) is upper bound on complexity. It may or may not be asymptotically tight; the bound $2m^2 \in O(m^2)$ is asymptotically tight but $2m \in O(m^2)$ is not, then we use o (little oh) notation to indicate an upper bound that is not asymptotically tight

o(g (m)) = {f(m): for some positive constant c > 0, $\exists$ a value m_0 > 0

So $0 \le f(m) < cg(m)$ $\forall m \ge m_0$

ω **Notation:** For lower bound, we use ω notation that is not asymptotically tight

ω(g(m)) = {f(m): for any c > 0, $\exists$ a constant m_0 s.t.

$0 \le cg(m) < f(m)$ $\forall m \ge m_0$

$m^2/2 \in \omega(m)$ but $m^2/2 \notin \omega(m^2)$

If $\lim_{m \to \infty} \frac{t(m)}{g(m)} = \infty$ that is, t(m) belongs to Omega of g(n).

Theorem 1.3: If $F(x) = b_m x^m + b_{m-1} x^{m-1} + \text{------------} + b_1 x + b_0$ where b_0, b_1, ------------ b_n are real numbers, then $f(x) \in O(x^m)$ if x > 1.

Proof: We know that $|x + y| \le |x| + |y|$ $\forall$ x, y $\in$ z

$F(x) = b_m x^m + b_{m-1} x^{m-1} + \text{------------} + b_1 x + b_0$

$|F(x)| = |b_m x^m + b_{m-1} x^{m-1} + \text{------------} + b_1 x + b_0|$

$\le |b_m| x^m + |b_{m-1}| x^{m-1} + \text{------------} + |b_1| x + |b_0|$

$\le x^m \{|b_m| + |b_{m-1}|/x^m + \text{------------} + |b_0|/x^m\}$

$|F(x)| \le x^m (|b_m| + |b_{m-1}| + \text{------------} + |b_0|)$

$\Rightarrow |F(x)| \le c\ x^m$ for x > 1

$\Rightarrow F(x) \in O(x^m)$

Question: Give big O estimation for function $f(x) = 3n \log n! + (n^2 + 5)\log n$

Solution:

First, $3n \log n!$ $\because \log n! \in O(n \log n)$

$3n \in O(n)$

$\Rightarrow 3n \log n! \in O(n^2 \log n)$

Again $(n^2 + 5)\log n$, $n^2 + 5 \le 3\ n^2$ where $n \ge 2$

$n^2 + 5 \in O(n^2)$

$\log n = O(\log n)$

$\Rightarrow (n^2 + 5)\log n \in O(n^2 \log n)$

$\Rightarrow$ And so, $3n \log n! + (n^2 + 5)\log n \in O(n^2 \log n)$

1.6 Recurrence relation

Consider a sequence S = {s_0, s_1, s_2 . . . . up to infinite}, a *recurrence relation* for S is an equation in which the general term s_n can be defined in terms of its previous terms s_0, s_1 . . . s_{n-1}. In Mathematics, a *recurrence relation* is a condition that recursively characterizes a succession. Each term of the grouping is characterized as a component of the former terms.

Example 1.5:
(a) $s_n - c s_{n-1} + d s_{n-2} = 3.2^n$ (in homogenous linear recurrence relation with constant coefficient),
(b) $s_n = c\, a(n/d) + f(n)$ (recurrence relation in divide and conquer (DAC) form),
(c) $s_n^2 - c\, s_{n-1} + d\, s_{n-2} = 3.2^n$ (nonlinear relation),
(d) $F_n = F_{n-1} + F_{n-2}$ (Fibonacci relation).

Solving a recurrence relation means to identify the sequence for which the recurrence relation exists.

1.6.1 Linear recurrence relation

A recurrence relation in form

$$a_n + b_1 a_{n-1} + b_2 a_{n-2} + - - - - - - - + b_k a_{n-k} = f(n), \ n \geq k \tag{1.1}$$

where b_1, b_2, ... b_k are all constants and f(n) is some function of n, known as linear recurrence relation with constant coefficient. If f(n) = 0, then the relation is identified as homogenous recurrence relation, otherwise nonhomogeneous recurrences relation. The degree of recurrence relation (1.1) is k, a_n is (n + 1)th term of the sequence.

1.6.2 Solution of homogeneous recurrence relation

(a) Characteristic equation method
The characteristic equation method is developed from the generating function method. For sequence $\{a_n\}_0^\infty$, the generating function is A(x), then we have

$$A(x) = \sum_{n=0}^{n=\infty} a_n x^n$$

Further, for a given recurrence relation, we can express A(x) as

$$A(x) = N(x)/D(x).$$

This D(x) is an important expression, for recurrence relation

$$a_n + b_1 a_{n-1} + b_2 a_{n-2} + \cdots\cdots + b_k a_{n-k} = 0$$

$$D(x) \text{ will be } (1 + b_1 x + b_2 x^2 + \cdots\cdots + b_k x^k)$$

Then characteristic polynomial will be

$$C(t) = t^k . \{D(1/t)\}$$

The characteristic equation will be C (t) = 0.

That is,

$$t^k + b_1 t^{k-1} + b_2 t^{k-2} + \ldots\ldots + b_k = 0$$

There two cases will arise:

(1) *When the roots are all distinct*

Let the roots be $\alpha_1, \alpha_2, \alpha_3, \ldots, \alpha_k$, then solution of recurrence relation will be

$$a_n = A_1 \alpha_1{}^n + A_2 \alpha_2{}^n + A_3 \alpha_3{}^n + \ldots + A_n \alpha_k{}^n$$

(2) *When the roots are not distinct*

Let the roots be $\alpha_1, \alpha_2, \alpha_3, \ldots\ldots, \alpha_n$ with multiplicity

$$m_1, m_2, \ldots\ldots m_i, \ldots, m_n, \quad m_1 + m_2 + \ldots + m_n = k.$$

Then the solution

$$a_n = \ldots + (A_0 + A_1 n + A_2 n^2 + \ldots + A_{m_i - 1} n^{m_i - 1} \alpha_i^n + \ldots, \text{ where } m_i \text{ is multiplicity of root } \alpha_i$$

Example 1.6: Find the general solution of recurrence relation

$$a_n - 3a_{n-1} + 2a_{n-2} = 0 \text{ for } n \geq 2, \text{ where } a_0 = 1 \text{ and } a_1 = 2.$$

Solution:

The characteristic equation is $x^2 - 3x + 2 = 0$, solving this equation

$$x = 2, 1$$
$$a_n = A_1 2^n + A_2 1^n$$

The starting conditions are $a_0 = 1$ and $a_1 = 2$

By putting n = 0 and 1 in eq. (1.2) and solving the equations, we get

$$A_1 = 1 \text{ and } A_2 = 0$$

As a result, the solution to the recurrence relation will be

$$A_n = 1.2^n$$

Example 1.7: Find the solution of the following relation

$$a_n - 6a_{n-1} - 9a_{n-2} = 0 \text{ with } a_0 = 1 \text{ and } a_1 = 6.$$

Solution:

The characteristic equation will be

$$t^2 - 6t - 9 = 0$$

Roots of the equation are 3 and 3.

Hence, the solution to the recurrence relation is

$$a_n = A_1 \cdot 3^n + A_2 \cdot n \cdot 3^n \text{ for some constants } A_1 \text{ and } A_2$$

To match the initial condition, we need

$$a_0 = 1 = A_1$$

$$a_1 = 6 = A_1 \times 3 + A_2 \times 3$$

Solving these equations yields $A_1 = 1$ and $A_2 = 1$.

Consequently, the overall solution is given by

$$a_n = 3^n + n \cdot 3^n = (1+n) \cdot 3^n$$

1.6.3 Solution to nonhomogeneous linear recurrence relation

A recurrence relation of the form

$$a_n + b_1 a_{n-1} + b_2 a_{n-2} + \cdots \cdots + b_k a_{n-k} = f(n), \, n \geq k$$

where $f(n) \neq 0$ is said to be a nonhomogeneous recurrence relation.

Consider the following relation:

$$a_n - a_{n-1} - a_{n-2} = n + 3$$

$$\text{LHS} \qquad\qquad \text{RHS}$$

where $f(n) = n + 3$.

LHS of the relation is the homogeneous part of the relation; so it is called the **associated linear homogeneous recurrence relation**. RHS is the second member of relation. To solve such relation, LHS is solved first and then RHS is solved to find a particular solution (PS). The results of both the sections are combined to form a complete solution of relation.

Finding the PS

There is no universal process to find a PS of a recurrence relation. Though if RHS of the relation takes a simple form, the method of inspection (or guessing) can be used. For example,

i.   If $f(n) = C$, a constant. Then, we suppose that the PS is also a constant A.

$$a_n + C_1 a_{n-1} + \cdots\cdots + C_k a_{n-k} = D \cdot a^n$$

ii. If $f(n)$ is in the exponential form, that is, a^n, then the trial solution for PS $= A \cdot a^n$, if a is not the root of a characteristic equation.

iii. If $f(n)$ is in the exponential form, that is, a^n, then the trial solution for PS $= A \cdot n^m a^n$, when a is root of a characteristic equation with multiplicity "m."

Example 1.10: Find the solution for the following nonhomogeneous recurrence relation,

$$a_n - 6a_{n-1} - 9a_{n-2} = 4.2^n.$$

Solution:
The roots of the equation $t^2 - 6t - 9 = 0$ are 3 and 3.

Hence, the solution of associated homogenous part is

$$a_n{}^H = (A_1 + A_2.n).3^n, A_1 \text{ and } A_2 \text{ are constants}$$

To compute the PS, we have $a = 2$, and a is not the root of equation. Then according to case (iii), the trial solution will be

$$a_n{}^P = A.2^n$$

To determine the value of A, we substitute $a_n = A.2^n$ in the given recurrence relation.

Example 1.11: What is the solution of the recurrence relation?

$$a_n - 7a_{n-1} + 10\, a_{n-2} = 7.5^n$$

Solution:
The roots of the equation $t^2 - 7t + 10 = 0$ are 2 and 5.

Hence, the solution of the associated homogenous part is

$$a_n{}^H = (A_1.2^n + A_2.5^n), A_1 \text{ and } A_2 \text{ are constants}$$

To compute the PS, we have $a = 5$, where a is the root of equation with multiplicity 1.

Then according to case (ii), the trial solution will be

$$a_n{}^P = A.n5^n$$

To determine the value of A, substitute $a_n = A.n5^n$ in the given recurrence relation:

$$An5^n - 7A(n-1)5^{n-1} + 10A(n-2)5^{n-2} = 7.5^n$$

$$A5^{n-2}\{25n - 7(n-1).5 + 10(n-2)\} = 7.5^2.5^{n-2}$$

$$A\{25n - 35n + 35 + 10n - 20\} = 7.25$$

$$15A = 7 \times 25$$

$$A = \frac{7 \times 5}{3}$$

$$A = \frac{35}{3}$$

$$P.S = \frac{35}{3}.n.5^n$$

$$a_n = a_n^H + PS$$

$$a_n = (A_1 2^n + A_2 5^n) + \frac{35}{3}n.5^n$$

1.7 Divide and conquer relations

Assume that an algorithm divides a problem (contribution) of size n into subproblems, where each subproblem is of size n/b. At that point, these subproblems are unraveled thusly, and the answers for these subproblems are joined to get the arrangement of the first problems. Assume that g(n) tasks are performed for such a division of a problem and join the subresults. At that point, if f(n) speaks to the quantity of activities required to tackle the problems, it follows that the function f fulfills the repeat connection

$$f(n) = c.f(n/b) + g(n).$$

This is called a **DAC recurrence relation.**

Example 1.12: The algorithm of binary search decreases the search for a component in a hunt arrangement of size n to the twofold search for this component in a pursuit arrangement of size n/2 (if n is even).
(i) For Binary Search,
$$f(n) = f(n/2) + A$$
(ii) The recurrence relation for merge sort,
$$T(n) = 2T(n/2) + cn, \text{ when } n > 1 \text{ and } A, \text{ when } n = 1 \text{ is a DAC relation.}$$

1.7.1 Solution to DAC recurrence relation

Example 1.13:
$$f(n) = 3f(n/2) + n$$

taking
$$n = 2^k, k = \log_2^n$$

let
$$f(n) = f(2^k) = b_k$$
$$f(n/2) = f(2^k/2) = f(2^{k-1}) = b_{k-1}$$

So recurrence relation is
$$b_k = 3b_{k-1} + 2^k$$

$$b_k - 3b_{k-1} = 2^k$$

The characteristic equation for recurrence relation will be

$$t - 3 = 0, \text{ so } t = 3$$

The homogeneous solution will be, $b_k^H = A.3^k$

Because a = 2 and 2 is not the root of characteristic equation, then according to case (ii), the trial solution for PS will be $B.2^k$, by substituting $b_k = B.2^k$ in the recurrence relation.

We have

$$B2^k - 3B2^{k-1} = 2^k$$

$$(2^{k-1})(2B - 3B) = 2.2^{k-1}$$

$$-B = 2, B = -2$$

So, $PS = B2^k = -2.2^k$.

So solution will be

$$b_k = A3^k + (-2)2^k$$

$$b_k = A.3^k - 2.2^k$$

$$b_k = a_n \Rightarrow a_n = A.3^{\log_2^n} - 2.n$$

$$a_n = A_0 n^{\log_2 3} - 2n$$

Let $3^{\log_2^n} = n^p$

Taking log of both sides at base 2

$$\log_2^n \log_2^3 = p.\log_2^n$$

$$p = \log_2^3$$

1.7.2 The recursion tree method

Example 1.14: Let the recurrence relation be $T(n) = 2T\left(\frac{n}{2}\right) + n$

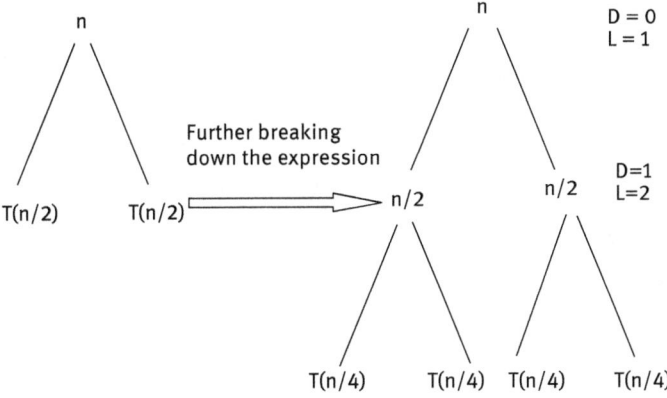

Figure 1.6: Tree representation of function T(n) of example 1.14.

To find the value of T(n), we need to find the sum of tree nodes (see figure 1.6) level by level (up to k levels). Let

$$\left(\frac{n}{2^k}\right) = 1, \quad \text{so}$$

$$n = 2^k, \quad \text{and} \quad k = \log_2 n$$

So

$$T(n) = \left\{ n + 2\frac{n}{2} + 4\frac{n}{4} + \ldots\ldots + n \right\} k \text{ times} + 2^k T(1)$$

$$T(n) = kn + 2^k \cdot c$$

$$T(n) = n\log_2 n + n \cdot c$$

$$\Rightarrow T(n) \in O(n\log_2 n)$$

Example 1.15: For relation $T(n) = 3\,T(n/4) + n$ obtain the asymptotic bound.
Solution: $T(n) = 3\,T(n/4) + n$.

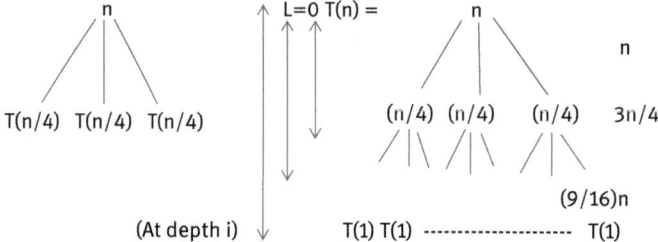

Figure 1.7: Tree representation of function $T(n)$ of example 1.15.

The subproblem size is $n/4^i$ at depth i. Thus, the subproblem size is 1 at last level (see figure 1.7), that is, $n/4^i = 1$ when $i = \log_4 n$

Thus, the tree has $(\log_4 n + 1)$ levels.

So we have

$$T(n) = n + (3/4)n + (3/4)^2 n + (3/4)^{\log_4 n \ 1} \cdot n + 3^{\log_4 n} T(1)$$

$$= n\left\{ 1 + (3/4) + (3/4)^2 + \cdots\cdots + (3/4)^{\log_4 n - 1} \right\} + 3^{\log_4 n} T(1)$$

$$T(n) < n\left\{ 1 + (3/4) + (3/4)^2 + -------- \right\} + 3^{\log_4 n} T(1)$$

$$T(n) < (1/(1-3/4))n + n^{\log_4 3} T(1)$$

$$T(n) < 4n + n^{\log_4 3} T(1)$$

$$T(n) \in O(n)$$

1.7.3 The substitution method

This technique is pertinent when we have floor or ceiling sign in the recurrence relation. The fundamental strides of substitution technique are:

1) Guess the type of the solution.
2) Use scientific enlistment to discover the constants and show that the solution works.

Example 1.16: $T(n) = 2T\left(\left\lfloor \frac{n}{2} \right\rfloor\right) + n$

Solution:

We start by concentrating on discovering the upper limit for the arrangement. We realize that floor and ceiling are generally deficient in tackling recurrence relation.

Therefore, we are creating recursion tree for the recurrence relation (see figure 1.8):

$$T(n) = 2T\left(\frac{n}{2}\right) + n$$

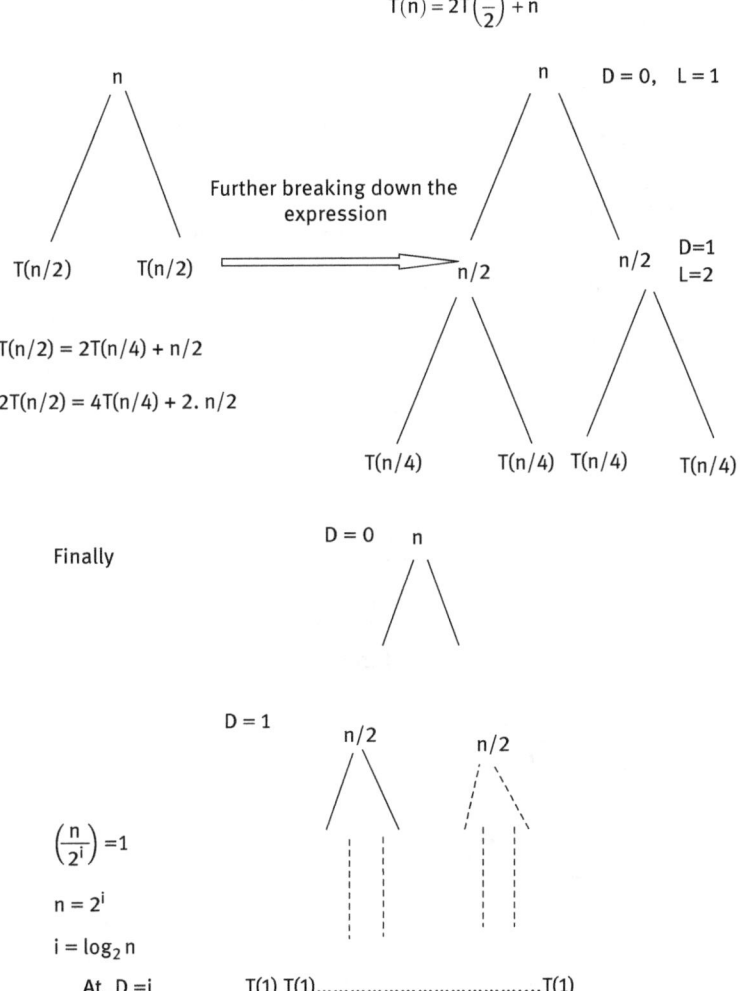

Figure 1.8: Tree representation of function T(n) of example 1.16.

So

$$T(n) = \left\{ n + 2\frac{n}{2} + 4\frac{n}{4} + \ldots\ldots\ldots + n \right\} i + 2^i T(1)$$

$$T(n) = in + 2^i \cdot c$$

$$T(n) = n\log_2 n + n.c$$

$$\Rightarrow T(n) \in O(n\log n)$$

Recurrence relation is $T(n) = 2T\left(\lfloor \frac{n}{2} \rfloor\right) + n$

Let $T(1) = 1$

we have, $T(n) \le c\, n\, \log n$

then we have $T(1) \le 0$

So, base case of inductive proof fails to hold.

Taking advantage of the asymptotic notation that $T(n) \le c\, n\, \log_2 n$ whenever $n \ge n$

$$T(2) = 4,\ T(3) = 5,\ \{\text{ for } T(1) = 1\}$$

Because assuming that $T(n) \le c\, n \log n$ holds for $\lfloor n/2 \rfloor$, that is

$$T\left(\lfloor n/2 \rfloor\right) \le c\, \left(\lfloor n/2 \rfloor\right) \log \left(\lfloor n/2 \rfloor\right)$$

Substitution method:

$$T(n) = 2T\left(\left\lfloor \frac{n}{2} \right\rfloor\right) + n,\ \underline{T(1) = 1}$$

$$T(n) \le c\, n \log n \qquad\qquad c \leftarrow \text{guess}$$

Let us assume, it is true for $\left\lfloor \dfrac{n}{2} \right\rfloor$

$$T\left(\frac{n}{2}\right) \le c\left\lfloor \frac{n}{2} \right\rfloor \log \left\lfloor \frac{n}{2} \right\rfloor$$

$$T(n) \le 2\left(c\left\lfloor \frac{n}{2} \right\rfloor \log \left\lfloor \frac{n}{2} \right\rfloor + n\right.$$

$$\le c\, n \log \left\lfloor \frac{n}{2} \right\rfloor + n$$

$$= c\, n\log n - c\, n\log 2 + n$$

$$= c\, n\log n - c\, n + n$$

$$= c\, n\log n - (c - 1)n$$

$$T(n) \le c\, n \log n$$

$$\text{for } c \ge 1$$

$$T(1) = 0 \text{ but } T(1) = 1$$

The base case for inductive proof fails to hold. When $(n > 3)$, the recurrence does not depend on $T(1)$ directly. Hence, setting new base of the inductive proof $(n = 2 \text{ and } n = 3)$

taking $c \geq 2$

$T(2) \leq c \, 2 \log 2$

$T(3) \leq c \, 3 \log 3$

$$T(2) = 4$$
$$T(3) = 5$$
$$T(4) = 10$$
$$= c \, 4 \log_2 4$$
$$= 8.c. \geq 16$$

$T(n) \leq c \, n \log_n$ for $c \geq 1$ Inductive proof for this

$T(2) = 4, T(3) = 5$

taking c large enough so

that $T(2) \leq 2c \, \log_2 2$

$T(3) \leq 3 \, c \log_2 3$

$$\text{taking} c \geq 2.$$

With $c \geq 2$, the solution works for $n > 1$.

1.7.4 Change of variable method

This method is applicable in some special cases. We will give an example.

Example 1.17:

$$T(m) = 2T(\lfloor \sqrt{m} \rfloor) + \log m$$
$$m = 2^n, \Rightarrow \sqrt{m} = 2^{n/2} \text{ and } \log m = n$$
$$\text{So we have } T(m) = T(2^n) = S(n) \text{(let)}$$
$$s(n) = 2 \, s(n/2) + n$$
$$s(n) \in O(n \log n)$$
$$T(m) \in O(\log m \, \log(\log m))$$

1.8 Master's theorem

Let T(n) be the nondecreasing function that fulfills the recurrence relation

$$T(n) = aT\left(\frac{n}{b}\right) + f(n) \quad \text{for } n = b^k \quad k = 1, 2, 3\ldots\ldots\ldots,$$

$T(1) = c$ and where $a \geq 1, b \geq 2, c > 0$

if $f(n) = n^d = \theta(n^d)$, where $d \geq 0$. Then

$$T(n) \in \begin{cases} \theta(n^d) & \text{if } a < b^d \\ \theta(n^d \log n) & \text{if } a = b^d \\ \theta(n^{\log_b^a}) & \text{if } a > b^d \end{cases}$$

Note: Similar result for O and Ω notations too.

Proof: We have the recurrence relation

$$T(n) = aT\left(\frac{n}{b}\right) + f(n), \text{ and } T(1) = c$$

$$f(n) = n^d, n = b^k \Rightarrow k = \log_b^n$$

$$T(n) = aT\left(\frac{n}{b}\right) + f(n)$$

Applying backward substitution,

$$T(b^k) = a\, T(b^{k-1}) + f\,(b^k)$$

$$= a[a\, T(b^{k-2}) + f\,(b^{k-1})] + f\,(b^k)$$

$$= a^2\, T(b^{k-2}) + a\, f\,(b^{k-1}) + f\,(b^k)$$

$$= a^3\, T(b^{k-3}) + a^2\, f\,(b^{k-2}) + a\, f\,(b^{k-1}) + f\,(b^k)$$

$$= \text{-------}$$

$$= a^k\, T(1) + a^{k-1}\, f\,(b^1) + a^{k-2}\, f\,(b^2) + \ldots + a^o\, f\,(b^k)$$

$$= a^k[T(1) + \sum_{j=1}^{k} f(b^j)/a^j]$$

$$\begin{cases} k = \log_b n \\ a^k = a^{\log_b n} \\ = n^{\log_b a} \end{cases}$$

$$T(n) = n^{\log_b a}\left[T(1) + \sum_{j=1}^{\log_b^n} \left(\frac{b^{jd}}{a^j}\right)\right]$$

$$\begin{cases} \because f(n) = n^d \\ f(b^j) = b^{jd} \end{cases}$$

$$\text{So, } T(n) = n^{\log_b a}\left[T(1) + \sum_{j=1}^{\log_b^n}\left(\frac{b^{jd}}{a^j}\right)\right]$$

$$T(n) = n^{\log_b a}\left[T(1) + \sum_{j=1}^{\log_b^n}\left(\frac{b^d}{a}\right)^j\right]$$

$$\because \sum_{j=1}^{\log_b^n}\left(\frac{b^d}{a}\right)^j = \begin{cases}\left(b^d/a\right)\left\{\left\{\left(b^d/a\right)^{\log_b n} - 1\right\}\Big/\left(b^d/a - 1\right)\right\}, & \text{if } b^d \neq a \\ \log_b^n \text{ if } b^d = a\end{cases}$$

Condition:

if $a < b^d$

$$\sum_{j=1}^{\log_b^n}\left(b^d/a\right)^j = \left(b^d/a\right)\left\{\frac{\left(b^d/a\right)^{\log_b n} - 1}{\left(b^d/a\right) - 1}\right\} \in \theta\left(\left(b^d/a\right)^{\log_b n}\right)$$

$$T(n) = n^{\log_b^a}\theta((b^d/a)^{\log_b n})$$

$$T(n) = \theta(b^{d\log_b n})$$

$$T(n) = \theta(b^{\log_b n^d})$$

$$T(n) = \theta(n^d)$$

Condition:

if $a > b^d$ then $\left(b^d/a\right) < 1$. Therefore,

$$\sum_{j=1}^{\log_b^n}\left(b^d/a\right)^j = \left(b^d/a\right)\left\{\frac{\left(b^d/a\right)^{\log_b n} - 1}{\left(b^d/a\right) - 1}\right\} \in \theta(1)$$

constant

Hence,

$$T(n) = n^{\log_b a}\left[T(1) + \sum_{j=1}^{\log_b^n}\left(b^d/a\right)^j\right] \in \theta(n^{\log_b a})$$

Again if $a = b^d \Rightarrow b^d/a = 1$

$$T(n) = n^{\log_b a}\left[T(1) + \sum_{j=1}^{\log_b^n}(b^d/a)^j\right]$$

$$= n^{\log_b a}[T(1) + \log_b n] \qquad \{a = b^d\}$$

$$\in \theta\ n^{\log_b a} \cdot \log_b n = \theta\left\{n^{\log_b b^d}\log_b n\right\}$$

$$= \theta(n^d \cdot \log_b n)$$

Question: Solve $T(n) = 2T(n/2 + f(n)$ $f(n) = n^2$

Solution: $a = 2, b = 2, f(n) = n^2 = \theta(n^2)$

$$n = 2^k \Rightarrow k = \log_2^n 2 \qquad\qquad b^d = b^2 = 2^2 = 4$$

$$4 > 2 \quad b^d > a$$

$$\Rightarrow T(n) \in \theta(n^d)$$

$$\Rightarrow T(n) \in \theta(n^2)$$

Question: Solve $T(n) = 4T\left(\dfrac{n}{2}\right) + f(n)$ $f(n) = n^2$

Solution: $a = 4, b = 2, f(n) = n^2 = \theta(n^2)$

$$f(n) = n^2 = n^d \Rightarrow d = 2 \qquad\qquad b^d = 2^2 = 4$$

$$\Rightarrow T(n) \in \theta(n^d \log n) \qquad\qquad a = b^d$$

$$\Rightarrow T(h) \in \theta(n^2 \log n)$$

Problem set

1. What is an algorithm? What is the need to study algorithms?
2. Explain the algorithm design and analysis process with a neat diagram.
3. Define:
    a) Time efficiency
    b) Space efficiency
4. What is the order of growth?
5. What do you mean by time complexity? Explain various notations used to sig-
    nify their complexities.
6. Explain omega notation. What are the various terms involved in it? Explain
    with the help of an example.
7. Describe the concept of time complexity of an algorithm when it is computed.

8. Explain with the help of an example the application that needs algorithmic content at the application level and discuss the function of the algorithms involved.
9. Describe in detail about the function $n^3/500 - 50n^2 - 10n + 9$ in terms of Θ notation.
10. What is the process to amend in an algorithm so that a best-case running time can be achieved?
11. Describe with the help of an appropriate mathematical induction to demonstrate that when n is an exact power of 2, the solution of the recurrence

$$T(n) = \begin{cases} 2 & \text{if } n = 2 \\ 2T(n/2) + n & \text{if } n = 2^a, \text{ for } a > 1 \end{cases}$$

is $T(n) = n\log n$.

12. Associate the order of growth of $\log_2 n$ and $\sqrt{n}$.
13. Demonstrate that the solution to $T(n) = 2T(n/2 + 17) + n$ is $O(n \log n)$.

Chapter 2
Sorting techniques

2.1 Sorting

Sorting is a method to arrange a data in ascending or descending order of magnitude. Sorting has great importance in the field of algorithms. There are several sorting algorithms available, which can be categorized on the basis of their performance in terms of space and time requirement. Some important sorting techniques are discussed in the following sections.

2.2 Insertion sort

Insertion sort depends on the decrease–and-conquer strategy. It is a straightforward arranging technique that fabricates the last arranged cluster (or show) each thing in turn. It is substantially less effective on enormous records than more advanced algorithms, for example, heap sort, merge sort, or quick sort. Be that as it may, insertion sort gives a few preferences:

- Implementation is easy.
- Good for small number of elements.
- More productive by and by than the greater part of the other straightforward quadratic (i.e., $O(n^2)$) algorithms, for example, selection sort or bubble sort.
- Adaptive, for example, effective for informational collections that are now considerably arranged.
- Stable, for example does not change the overall order of components having equivalent keys or qualities.
- In-place sorting.
- Online, for example, can sort a rundown as it gets it.

At the point, when individuals physically sort something (for instance, a deck of playing a card game), most utilize a technique that is like insertion sort.

Algorithm: *Insertion Sort(Array, size)*

```
For n: =1 to (size-1) do
  {
    ele: = Array[size];
    j : = (size-1);
    While (j≥ 0) and Array[j] > ele) do
      {
      Array [j+1]: =Array [j];
```

https://doi.org/10.1515/9783110693607-002

```
    j: = j-1;
    }
    Array [j+1]: = ele;
  }
```

Example 2.1:

2.2.1 Analysis of insertion sort

The worst-case analysis: The nature of input drives the number of key comparisons performed in execution. If elements of array are in decreasing order, than it is worst input, that is,

$$A[0] > A[1], A[1] > A[2] \ldots \ldots \ldots A[m-2] > A[m-1]$$

Thus number of key comparisons is,

$$c_{worst}(m) = (m-1) + (m-2) + \ldots + 3 + 2 + 1$$

$$= \frac{m(m-1)}{2} \in \theta(m^2)$$

The best-case analysis: If elements of array are in increasing order, that is, already sorted in ascending order, then this is the best input.

Only one comparison per iteration

$$\text{Thus, } C_{best}(m) = \sum_{i=1}^{m-1} 1 = (m-1) \in \theta(m)$$

The average-case analysis: This is the case when array elements are randomly taken and insertion sort algorithm is applied. In average case, we are inserting element A[i] in subarray A[0:i-1], where average half of the element in subarray A[0:i-1] are less than A[i] and half of the element are greater than A[i] as shown in Figure 2.1.

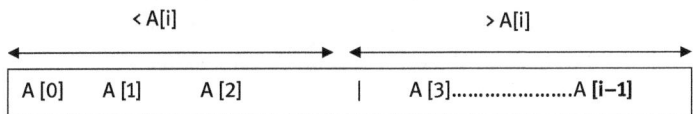

Figure 2.1: Partition about A[i].

On an average, we check half of the subarray A[0: i−1], so for i^{th} iteration, number of comparisons will be $c_i = \frac{i}{2}$

Thus, total number of comparisons,

$$C_{Avg}(m) = \sum_{i=1}^{m-1} \frac{i}{2} = \frac{1}{2}\left(\frac{m(m-1)}{2}\right)$$

$$= \frac{1}{4}(m^2 - m)$$

$$\approx \left(\frac{m^2}{4}\right) \approx \theta(m^2)$$

2.3 Quick sort

The fundamental variant of quick sort calculation was concocted by C. A. R. Hoare in 1960, which he officially presented later in 1962. It depends on the rule of divide-and-conquer (DAC) technique. It is an algorithm of decision and is large since it is not hard to execute. It is a decent "broadly useful" arranging calculation and expends generally less assets during execution.

Quick sort works by apportioning a given cluster Ar[p . . r] into two nonvoid sub-arrays Ar[p . . q−1] and Ar[q + 1 . . r] to such an extent that each key in Ar[p . . q−1] is not exactly or equivalent to Ar[q], which thus not exactly or equivalent to each key in Ar[q + 1 . . r]. We call this component (Ar[q]) "PIVOT." At that point, the two subarrays are arranged by recursive calls to quick sort. The specific situation of the segment relies upon the given array and index q is calculated as a piece of the partitioning methodology.

It is a significant arranging method that is dependent on DAC approach. Not at all like merge sort that separates the input as indicated by the situation in array; quick sort partitions them as per their values explicitly. It revamps elements of a given exhibit Ar[1:n] to its parcel, a circumstance where all the components before some segment s are smaller than or equivalent to Ar[s], and all the component after segment are more noteworthy than or equivalent to Ar[s]. Clearly, after parcel about s has been accomplished, Ar[s] will be in its last situation in the arranged array. And we can keep arranging the two subarrays of the components going before and following Ar[s], autonomously by the equivalent method. First, we select an element as for whose esteem we are going to partition the subarray. The most straightforward procedure of selection the "PIVOT" is to set first element of subarray as PIVOT element. This arranging strategy is considered as in place since it utilizes no other array.

2.3.1 DAC approach

Divide: Split array Ar [1: size] into Ar [1: d−1] and Ar [d + 1: size] where d is determined as part of division.
Conquer: Sort Ar [1: d−1] and Ar [d + 1: size] recursively.
Combine: All this leaves sorted array in place.

Algorithm: *Partition (Ar, d, p)*
//This is partition of array Ar[d:p-1] about Ar[d].

```
{
      v: = Ar[d]; it: = d; j: = p;
      repeat
            {
                  repeat{
                        it : = it +1;
                        }until (Ar[it] ≥ v);
                  repeat {
                        j: = j-1;
                        }until (Ar[j] ≤ v);
                  if (it < j), then interchange(Ar, it, j);
            }until (it ≥j);
      Ar[d]: = Ar[j]; Ar[j]: = v;
      return j;
}
```

Algorithm: *Interchange (Ar, it, j)*
// Interchange value Ar[it] and Ar[j]

```
{
      temp:= Ar[it];
      Ar[it]: = Ar[j];
      Ar[j]: = temp;
}
```

Algorithm: *Quick Sort (m, n)*

```
{
      If (m<n) then
                  {                     // break the problem into sub problems
                       index:= Partition(Ar, m, n);
                       Quick Sort (m, index-1);
                       Quick Sort (index+1, n);
                  }
}
```

Example 2.2:

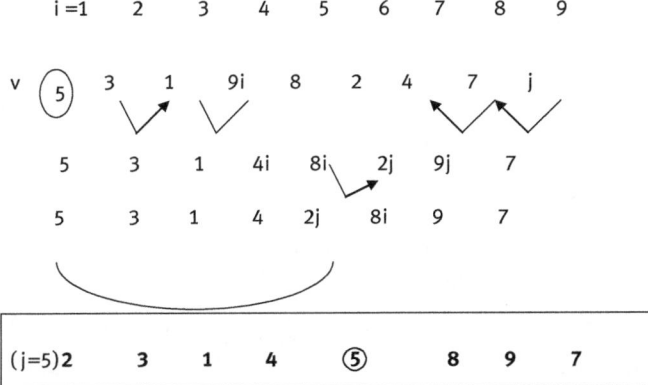

Figure 2.2: Example of Quick Sort.

2.3.2 Analysis of quick sort

The worst-case analysis: In most worst-case scenarios, one of the two subarrays will be vacant while since of the other will be only one less than the size of subarray being partitioned. This circumstance will happen when the given array is already arranged. In that case, first element is taken as PIVOT element. The left to right output will stop on A[2] while option to left sweep will go as far as possible and reach A[1], demonstrating the split at position 1. So after (n + 1) comparisons, we will get this segment and exchanging the PIVOT A[1] to itself. The algorithm will wind up with the carefully increasing array A[2: n] to sort. This procedure of carefully increasing array of lessening size will proceed until the last one A[n-1:n] needs to be prepared. So the total numbers of key comparisons made will be

$$C_{worst} = (n + 1) + n + (n - 1) + \ldots\ldots\ldots + 3$$

$$= \left\{ \frac{(n + 1)(n + 2)}{2} - 3 \right\}$$

$$\Rightarrow C_{worst}(n) \in \theta(n^2)$$

The best-case analysis: If all the split happen in the middle of corresponding subarray, we will still have the best case. The numbers of key comparisons will be:

$$C_{best}(n) = 2C_{best}\left(\frac{n}{2}\right) + cn, \text{for } n > 1$$

$$C_{best}(n) = 0, \text{for } n = 1 \tag{2.1}$$

solving this by taking $n = 2^k$, we get

$$C_{best}(n) \in \theta(n\log_2 n)$$

Note: *It is noted that partition is linear in size, that is, on each level, it is $\theta(n)$. Thus, we get recurrence relation (2.1). For a split only number of comparison are $\leq n$ (best case).*

The average-case analysis: The input size is n and we assume that partition split can happen at any position s (for $1 \leq s \leq n$) with same probability (1/n). For the extreme case, when split take place at first position, the number of element comparisons are (n + 1) then we have

$$C_A(n) = (n + 1) + \frac{1}{n}\left\{ \sum_{k=1}^{n} [C_A(k - 1) + C_A(n - k)] \right\}\ldots\ldots \tag{2.2}$$

$$\left\{ \begin{array}{l} \text{Expected value of a random variable} \\ E(x) = p_1 x_1 + p_2 x_2 + \ldots\ldots\ldots + p_n x_n \end{array} \right.$$

$$\left. \begin{array}{l} x_i \rightarrow \text{value of random variate} \\ p_i \rightarrow \text{its probability} \end{array} \right\}$$

With condition $C_A(0) = 0$ and $C_A(1) = 0$. Equation (2.2) is a recurrence relation and for the first partition only which may take place at any point (position) for different value of k. Multiplying both side by n,

$$nC_A(n) = n(n+1) + 2[C_A(0) + C_A(1) + \ldots\ldots C_A(n-1)]\ldots\ldots\ldots \tag{2.3}$$

Replacing n by (n–1) in (2.3), we have

$$(n-1)C_A(n-1) = n(n-1) + 2[C_A(0) + \ldots\ldots + C_A(n-2)]\ldots\ldots \tag{2.4}$$

Subtracting (2.4) form (3), we have

$$nC_A(n) - (n-1)C_A(n-1) = 2n + 2C_A(n-1)$$

$$\text{or } \frac{C_A(n)}{(n+1)} = \frac{C_A(n-1)}{(n)} + \frac{2}{(n+1)} \ldots\ldots\ldots\ldots\ldots\ldots\ldots\ldots\ldots \tag{2.5}$$

Repeatedly using the equation for $C_A(n-1)$, $C_A(n-2)$, ..., we get

$$\frac{C_A(n)}{(n+1)} = \frac{C_A(n-2)}{(n-1)} + \frac{2}{(n)} + \frac{2}{(n+1)}$$

$$= \frac{C_A(n-3)}{(n-2)} + \frac{2}{(n-1)} + \frac{2}{(n)} + \frac{2}{(n+1)}$$

$$= \frac{C_A(1)}{2} + 2\sum_{k=3}^{n+1}\left(\frac{1}{k}\right)\{Q\, C_A(1) = 0\}$$

$$\Rightarrow C_A(n) \le 2(n+1)[\log_e(n+1) - \log_e 2] \quad \left\{\sum_{k=3}^{n+1}\left[\left(\frac{1}{k}\right) \le \int_2^{n+1} 1/x\, dx = [\log_e(n+1) - \log_e 2]\right]\right.$$

$$\Rightarrow C_A(n) \le 2(n+1)[\log_e(n+1) - \log_e 2]$$

$$\Rightarrow C_A(n) \le O(n\log_e n)$$

or $C_A(n) \in O(n\log_2 n)$

$n\log_e n = cn\log_2 n$, where c is a specific constant

and $2n\log_e n \approx 1.38n\log_2 n\}$

2.4 Merge sort

Merge sort is an O(n log n) correlation-based arranging algorithm. Most executions produce a steady sort, implying that the usage protects the input order of equivalent elements in the arranged array. It is based on DAC strategy. Merge sort was proposed by John von Neumann in 1945. It sorts a given array Ar[1:n] by dividing it two halves Ar[1:⌊n/2⌋] and Ar[⌊n/2⌋ + 1: n], sorts every one of them recursively and afterward consolidates the two smaller sorted array into a single sorted one.

Algorithm: *Merge Sort (low, high)*

```
{
If(low<high) Then
    {
    mid:=⌊(low+high)/2⌋
    MergeSort(low,mid);
    MergeSort(mid+1 ,high);
    Merge(low ,mid ,high);
    }
}
```

Algorithm: *Merge (low, mid, high)*
// AU[ ] is an auxiliary global array.

```
{
    h:=low; x:=low; y:=mid+1;
    while (x ≤ mid) and (y ≤ high)do
    {
      if(Ar[x]≤ Ar[y] Then);
        {
          Au[h]: = Ar[x];
          x: = x+1;
        }
      else
        {
          Au[h]: = Ar[y];
          y: = y+1;
        }
        h: = h+1;
    }
if(x> mid) Then
  {
    for z: = y to high do
      {
        Au[h]: = Ar[z];
        h: = h+1;
      }
else
    for z: = x to mid do
      }
      Au[h]: = Ar[z];
      h: = h+1;
  }
```

```
for z: = low to high do
  Ar[z]: = Au[z];
}
```

Example 2.3: For example,

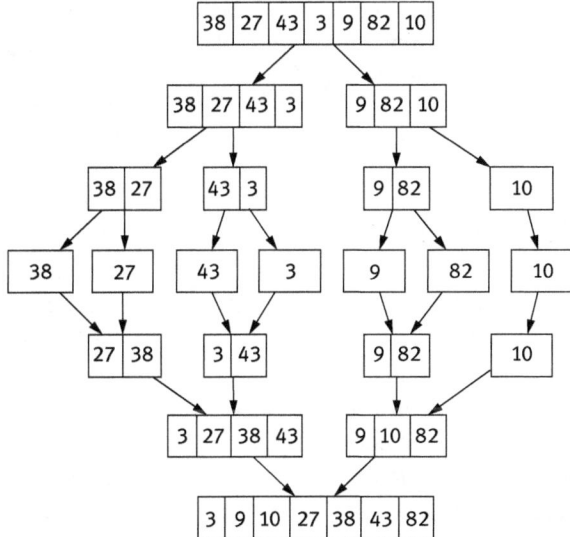

Figure 2.3: Example of Merge Sort.

2.4.1 Analysis of merge sort

The running time of an algorithm can be described by recurrence relation when the algorithm contains a recursive call to itself. For **MergeSort()**, we assume for simplicity that $n = 2^k$. A recurrence for DAC algorithm is based on three steps.

– **Divide:** This step finds the middle of subarray, so taking a constant time only.
– **Conquer:** Two n\2-sized subproblems are solved recursively, which adds $2T(n/2)$ to the running time.
– **Combine:** Merging process (on n element) takes time proportional to number of element, that is, cn.

$$So\ T(n) = \begin{cases} 1, & n = 1 \\ 2T\left(\frac{n}{2}\right) + cn, & n > 1 \end{cases}$$

where $n = 2^k$

Solving by substitution method,

$\Rightarrow T(n) \in \theta(n \log_2 n)$

The complexity of merge sort is of $O(n\log_2 n)$. Merge sort is not in place sorting method as it requires an auxiliary array to support the sorting. Also, this algorithm is not space efficient since it requires extra array space.

2.5 Heap sort

Heap sort is a sorting algorithm that is based on comparisons. Heap sort is considered to be improved selection sort, and like selection sort, it isolates its input into a sorted and an unsorted locale, and it iteratively recoils the unsorted locale with the aid of extricating the littlest factor and transferring that to the sorted locale. The advancement accommodates of the use of a heap structure in preference to a linear one so that the time to find the smallest is least. It is an in-place but not stable algorithm. It was proposed by J.W.J. Williams in 1964 and became the beginning of the heap.

Complete binary tree: It is a binary tree for which the vertex level order indices of vertices forms a complete interval 1, 2, 3, ..., n of integers (see Figure 2.4).

Heap: A heap is a complete binary tree with the property that the value at each node is at least as large as the values at its children (see Figure 2.5). This property is called heap property and this heap is called *max heap*. In contrast to this, in *min heap*, the value is as small as its children.

Vertex level order index: If p is the root then index (p) = 1, and if lc is left child of vertex p, then index (lc) = (index p)*2. If rc is right child of same vertex p, then index (rc) = 1 + (index p)*2.

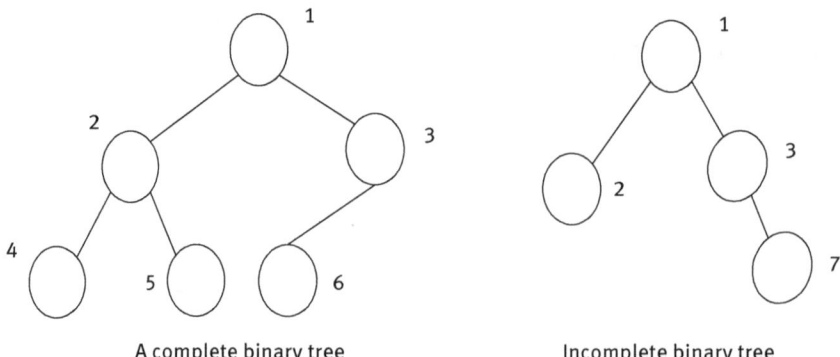

A complete binary tree   Incomplete binary tree

Figure 2.4: Complete and Incomplete Binary Tree.

The minimum height of a binary tree on n nodes is$\lceil [\log_2(n + 1)] \rceil$.

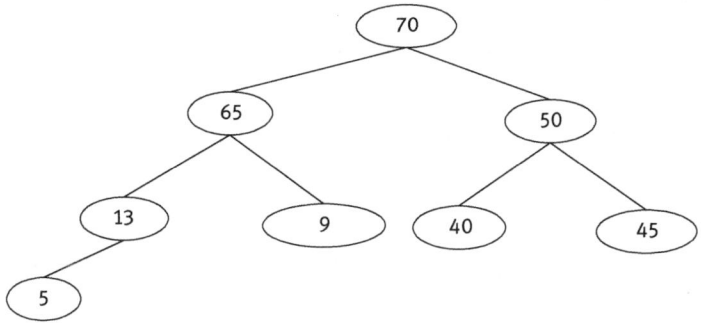

Figure 2.5: Example of Max Heap.

Adjust() is a function take three inputs; array Ar[], integer k, and m. It take Ar[1:m] as a complete binary tree. If subtree rooted at 2k and 2k + 1 are already max heap, then adjust will rearrange element of Ar[] such that the tree rooted at k is also max heap. For m element in Ar[1:m], we can create heap by applying *Adjust()*, since the leaf nodes are already heaps. So, we can begin by calling *Adjust()* for the parents of leaf nodes and then work up level by level until the root is reached.

Algorithm: *Adjust (Ar, k, m)*

```
{
      i: = 2k; el: = Ar[k];
      While (i ≤ m) do
            {
                  If(i <m) and (Ar[i] < Ar[i+1]) then
                        i:= i+1;
                  If(el ≥ Ar[i]) then break;
                  Ar[ Li/2J ] : = Ar[i]; i: = 2i;
            }
      Ar[i/2] : = el;
}
```

The function *Heapify (Ar,m)* readjust the element in Ar[1:m] to form a heap.

Algorithm: *Heapify (Ar, m)*

```
{
for k = ⌊m/2⌋ to 1step-1 do
Adjust(Ar, k, m);
}
```

Example 2.4: The elements of an array to be sorted are

20	35	32	42	30	50	46

Figure 2.6 shows sorting process of the array elements.

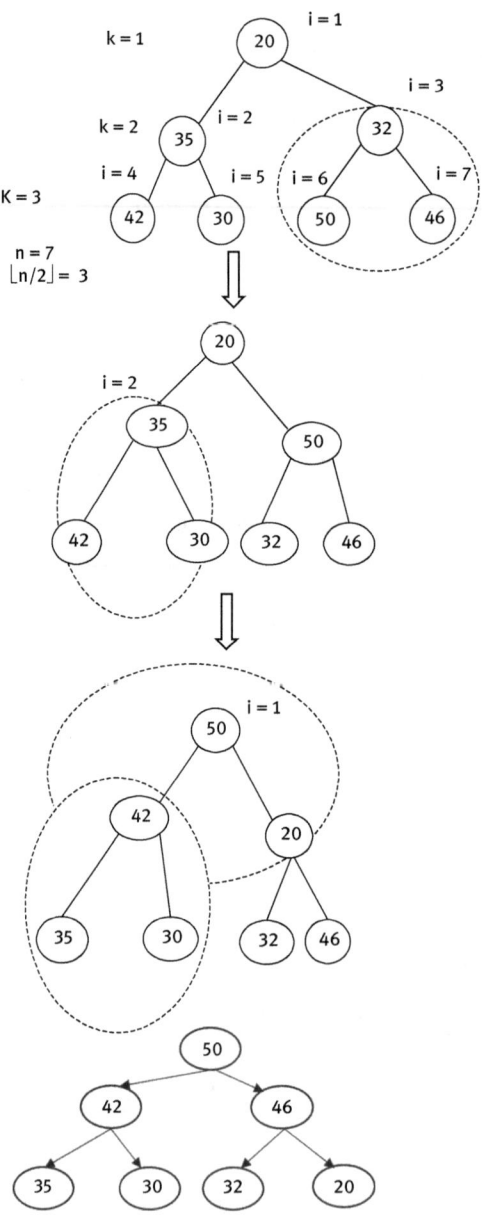

Figure 2.6: Process of Heap Sort.

2.5.1 Analysis of Heapify

For the worst case analysis of *Heapify*, let $2^{k-1} \leq n < 2^k$. The levels of the n node complete binary tree are numbered 1 to k. Number of iteration of *Adjust*, for level i, it is **(k-i)**. Number of nodes at level **i** are given by 2^{i-1}. So, the total time for Heapify is proportional to

$$\sum_{1 \leq i \leq k} 2^{i-1}(k-i) = \sum_{1 \leq i \leq k-1} i 2^{k-i-1} \leq n \sum_{1 \leq i \leq k-1} \frac{i}{2^i} \leq 2n = O(n)$$

So, *Heapify* takes a linear time to create the heap using *Adjust* function.

2.5.2 Heap sort

Algorithm: *HeapSort()*
```
{
   Heapify (Ar,m);                    // transforming array into max heap
   for k:= m to 2 step −1 do
   {
     Temp:= Ar[k];
     Ar[k]: = Ar[1];
     Ar[1]: = Temp;
     Adjust (Ar,1,k-1);
   }
}
```

2.5.3 Analysis of heap sort

Heapify requires O(n) operation, and *Adjust* possibly requires $O(\log_2 n)$, so the worst-case time complexity for *HeapSort* is $O(n \log_2 n)$.

2.6 Sorting in linear time

There are some sorting algorithms that run quicker than O(n logn) time but they require unique assumptions about the input collection to be sort. Here we present three linear time sorting algorithms; counting type, radix sort, and bucket type. Bucket sort may be used for a lot of the identical duties as counting kind, with a similar time analysis; however, compared to counting type, bucket type calls for related lists, dynamic arrays, or a big quantity of reallocated memory to hold the units of gadgets within each bucket, whereas counting type instead stores a single number (the count of elements) per bucket.

2.7 Counting sort

Counting sort and its application to radix sorting was both invented by Harold H. Seward in 1954. Counting sort is an algorithm for sorting a set of objects in step with keys which can be small integers, that is, it is for an integer sorting set of rules. It operates by counting the number of gadgets that have every wonderful key value and using mathematics on those counts to decide the positions of each key value in the output sequence. However, it is frequently used as a subroutine in another sorting algorithm, radix type, that could cope with large keys more efficiently. Because counting kind uses key values as indexes into an array, it is not a comparison type, and the $\Omega(n \log n)$ decrease certain for comparison sorting does not follow to it.

This sorting technique is based on frequency distribution. If values are integers between some lower bound l and upper bound u, we compute frequency of these values and store them in array F[0:u−1].

A =	15	12	15	16	19	36	15	15

l = 12, and u = 36
Value: 12, 15, 16, 19, 36
$\Sigma f_i = 8 =$ freq: 1, 4, 1, 1, 1
    Cumulative freq: 1, 5, 6, 7, 8

In this algorithm, we copy elements into a new array s[0:n−1]. The elements of array A whose values are equal to the lower possible value l are copied into the first F(0) elements of Š, that is, position 0 to F[0]−1, and the elements of value just greater then l are copied to position from F(0) to (F(0) + F(1)−1) and so on.

Algorithm: *Counting-Sort (Ar[0:size-1])*

```
{
    for g : =0 to (u-1) do
        D[g]: = 0:
    for j: = 0 to (size-1) do
        D[Ar[j]]: = D[Ar[j]]+1:
    for k: 1 to (u-1) do
        D[k]: = D[k-1]+D[k];
    for i: = (size-1) down to 0 do
        k:=Ar[i]-1
        S[D[k]-1] = Ar[i];
```

```
    D[k] : = D[k]-1;
  Return S;
}
```

Counting sort is a stable sorting algorithm as it copies elements in the final array in that order in which they are encountered in the unsorted array.

Example 2.5:

Input A =	6	9	7	6	4	5	6	7	9	5

l = 4, u = 9
u−l = 9−4 = 5,n = 10
Eement:4 5 6 7 8 9
Frequency: 1 2 3 2 0 2
J: = 1 to 5
Cumulative Frequency: 1 3 6 8 9 10
For i: (n-1) to 0 do   D[0] = 1
                                   D[1] = 3
                                   D[2] = 6:
                                   D[5] = 10
i = n-1(1st iteration)
n = 10
j = 5−4 = 1 { **A[9]** = 5}
S[3−1]: = 5
S[2]: = 5

	5		5						

D[1]: = 2(Remaining)
Finally, we will get sorted array.

2.7.1 Analysis of counting sort

Counting sort algorithm makes use of most effective simple for loops without any complex concept like recursion or subroutine calls, it is easy to analyze. The general time complexity of the counting sort is

$$\theta(k) + \theta(n) + \theta(k) + \theta(n) = \theta(n+k) = \theta(n). \text{ if } k \text{ is a constant}$$

2.8 Radix sort

Radix sort is applicable to the unit of n integers, and each integer has d digit. We start the sorting repetitively. We start with lowest order digit and finish at the highest order. The sorting is stable.

Algorithm: *RadixSort (Ar, dig)*
```
{
    for k = 1 to dig,
    Perform stable sort to sort array Ar on the digit k.
}
```

2.8.1 Analysis of radix sort

Let us assume that the stable sort runs in O(n + b). Then the running time of Radix sort will be O(dig(n + b)) = O(dig*n + dig*b). If dig is constant and b = O(n), the complexity is O(n).

Example 2.6: Input A = {677, 595, 295, 397}

677	59[5]	6[7]7	[2]95	295
595 ⇒	29[5] ⇒	5[7]5 ⇒	[3]97 ⇒	397
295	67[7]	2[9]5	[5]95	595
397	39[7]	3[9]7	[6]77	679

2.9 Bucket sort

Bucket sort is a linear time algorithm. Input is random and is distributed evenly in the interval [0, 1] the range is divided into n equal intervals. (see Figure 2.7). These intervals are called buckets. As the input is uniformly distributed, it is not expected that many numbers fall into one bucket. In bucket sort, first all the buckets are sorted individually and then concatenate them to get sorted array.

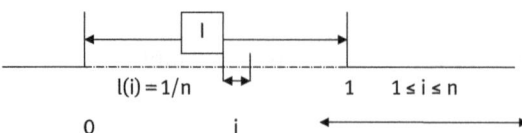

Figure 2.7: Uniform distribution.

Algorithm: *BucketSort(Ar)*

```
{
    size: = length[Ar];
    for k: 1 to size do
        insert Ar[k] into List Bu[ Lsize Ar [k] ⌋ ];
    for k: = 0 to (size-1) do
        sort bucket Bu[k] with insertion sort;
    Combine Bu[0], Bu[1],. . .. .. .. . .. . .. .Bu[size-1], in order.
}
```

To see how *BucketSort* works, consider two elements Ar[p] and Ar[q]. If both the elements fall in the same bucket, they follow comparative order in the sorted array because insertion sort arranges the buckets. Let us assume that these elements fall into diverse buckets, Bu[p'] and Bu[q'], respectively, and suppose without loss of generality that p' < q'. When the buckets are combined, Bu[p'] comes before Bu[q'], and therefore Ar[p] precedes Ar[q].

2.9.1 Analysis of bucket sort

If n_i be the random variable, it denotes the number of elements placed in Bucket Bu[i]. Except the sorting using insertion sort all instruction takes time of order $\theta(n)$ and there are n calls for insertion sort, so time for *BucketSort* is

$$T(n) = \theta(n) + \sum_{i=0}^{n-1} O(n_i^2)$$

In order to assess this summation, we have to find the distribution of each random variable n. We have n elements and n buckets. The probability of falling an element in a ith bucket is 1/n. This problem is the same as that of "Balls-and-Bin" problem. Hence, the probability follows the binomial distribution, which has

Mean: $E[n_i] = np = 1$

Variance: $Var[n_i] = np(1-p) = 1 - \dfrac{1}{n}$

Taking expectations of both side

$$E[T(n)] = \theta(n) + \sum_{i=0}^{n-1} E[O(n_i^2)] \ \{n_i \rightarrow \text{random variable, number of element in Bucket B[i]}\}$$

$$= \theta(n) + \sum_{i=0}^{n-1} O(2 - \dfrac{1}{n})$$

$$= \theta(n) + n.O(2 - \frac{1}{n})$$

$$= O(n)$$

Therefore, the entire *BucketSort* algorithm runs in linear expected time.

Example 2.7: Sort the array A using bucket sort
A = {0.23, 0.37, 0.15, 0.18, 0.34, 0.88, 0.78, 0.38, 0–99, 0.92}
N = 10, the elements 0.34, 0.37, 0.38 mapped at i = 3

$$n \times 0.34 = \lfloor 10 \times 0.34 \rfloor = 3$$
$$n \times 0.34 = \lfloor 10 \times 0.37 \rfloor = 3 \qquad \text{Buckets } n = 10$$
$$n \times 0.34 = \lfloor 10 \times 0.38 \rfloor = 3$$

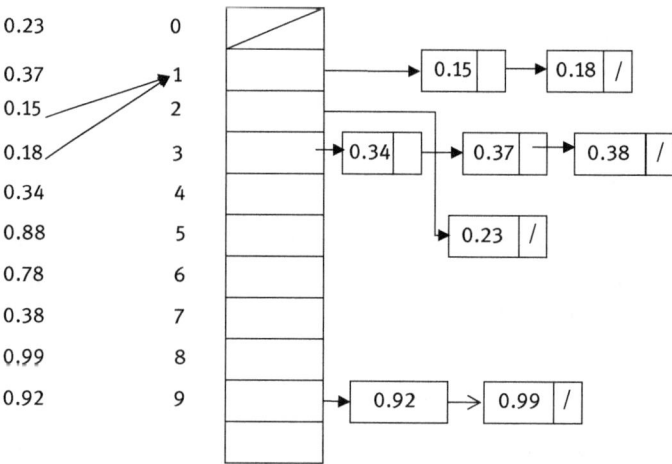

Figure 2.8: Example of Bucket Sort.

2.9.2 Binomial distribution

$$\text{Var}(n_i) = E(n_i{}^2) - \{[E(n_i)]\}^2$$

$$\text{Var}(x_i) = npq = (1 - \frac{1}{n})$$

$$E(n_i{}^2) = \text{Var}(n_i) + \{E(n_i)\}^2$$

$$= 1 - 1/n + 1 = 2 - 1/n$$

$$E(n_i) = np = n.1/n = 1$$

Problem set

1. What is meant by sorting?
2. List two different types of sorting?
3. Give the use of radix sorting.
4. Discuss the complexity of quick sort.
5. List and explain the sequence of steps required to sort numbers in the main memory using merge sort.
6. Give the differences between the insertion sort and selection sort.
7. Write an algorithm for the procedure MIN-HEAPIFY (A,i), which performs the corresponding manipulation on a min heap. How does the running time of MIN-HEAPIFY compare to that of MAX-HEAPIFY?
8. Explain the quick sort calculation with a model and furthermore draw the tree structure of the recursive calls made.
9. Analyze the productivity of quick sort calculation.
10. Explain the merge sort calculation with a model and furthermore draw the tree structure of the recursive calls made.
11. Analyze the effectiveness of merge sort calculation.
12. Give the insertion sort calculation and break down the effectiveness.
13. What is heap? What are the various kinds of heaps?
14. Explain how would you develop heap?
15. Explain the heap sort calculation with a model.
16. Explain the Horspool's calculation with a model.
17. Sort the following array
    90, 194, 245, 44, 23, 19, 60, 39, 76, 14, 1, 8, 54
    utilizing every one of the accompanying strategies:
    (i)   Merge sort
    (ii)   Quick sort
    (iii) Insertion sort
    (iv)   Selection sort
    (v)   Heap sort
    Further, tally the quantity of activities by each arranging technique.
18. Write an algorithm for quick sort for isolating the variety of components which is to be arranged and, furthermore, write the calculation for converging in the wake of arranging.

Chapter 3
Algorithm design techniques

This chapter presents techniques to design algorithms which are based on: greedy technique, dynamic programming, backtracking approach, and branch-and-bound method. Amortized analysis process is also discussed in this chapter.

3.1 Greedy approach

Greedy approach is perhaps the simplest of production techniques. This is usually used to solve optimization issues, such as finding the shortest path through a network from one node to the next and choosing the best order for s series of jobs to be performed on a computer. In greedy approach, we make the option that looks better in the current situation. Greedy algorithm works by choosing the most promising solution at any instant. It never reassesses this decision, whatsoever condition may arise later. It follows the concept of the **current best**.

3.1.1 Traveling salesman problem

In Traveling Salesman Problem, a salesperson has to visit n places or cities, in such a manner that all the cities must be visited only once and in the end, he returns to the city from where he starts, with minimum cost. The greedy approach uses the concept of **"Nearest neighbor"** and selects the city which is closest to the current city.

Example 3.1: The distance matrix for given complete graph G (refer figure 3.1).

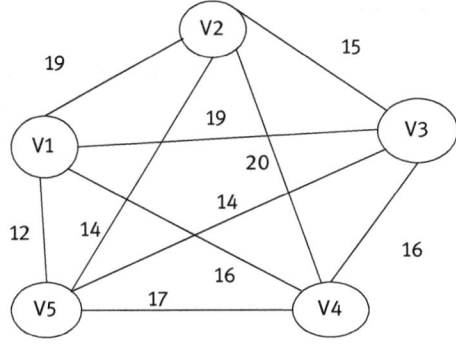

https://doi.org/10.1515/9783110693607-003

	V1	V2	V3	V4	V5
V1	0	19	19	16	12
V2	19	0	15	20	14
V3	19	15	0	16	14
V4	16	20	16	0	17
V5	12	14	14	17	0

Figure 3.1: A Graph G and its Distance Matrix.

V1-V5-V2-V3-V4-V1
Tour cost: −12 + 14 + 15 + 16 + 16 = 73,

3.1.2 Fractional knapsack problem

We have to place n objects in a given knapsack. The knapsack capacity is W. Weight of the ith object is w_i, if a fraction x_i $(0 \le x_i \le 1)$ of ith item is chosen to fill the knapsack then the profit earned is $v_i x_i$, where v_i is the value earned by placing ith item. The aim is to fill up the knapsack so that the profit will be maximum. Then linear programming problem (LPP) for above problem will be,

$$z = \sum_{1=1}^{n} v_i x_i$$

$$\text{such that } \sum_{i=1}^{n} w_i x_i \le W$$

$$\text{and } 0 \le x_i \le 1, \ 1 \le i \le n$$

Algorithm: *Greedy-Knapsack(W, n)*
// v [1: n] is the value array, w [1:n] is the weight array and x[1:n] contains the fractional value of objects placed in knapsack. The objects are sorted in increasing order s. t.

$$\frac{v[i]}{w[i]} \ge \frac{v[i+1]}{w[i+1]}$$

```
{
    for k : = 1 to n do
          x[k]: = 0.0;
    U: = W;
    for k: = 1 to n do
            {
                if (w[k] > U) Then break;
                x[k]:=1.0; U: = U−w[k];
```

```
    }
  if(k ≤ n) Then x[k]: = U/w[k];
}
```

The time complexity of the algorithm is O(n).

Example 3.2: Solving the given fractional knapsack problem with greedy approach (see table 3.1). Capacity of knapsack W = 16, Number of items is 4, the value and weights are given in the table.

Table 3.1: List of items to be filled in knapsack.

Item (i)	v_i	w_i	v_i/w_i
1	10	5	2
2	12	4	3
3	8	4	2
4	20	5	4

Arranging items in order $\frac{v[i]}{w[i]} \geq \frac{v[i+1]}{w[i+1]}$ (see table 3.2)

Table 3.2: Items arranged in order.

Item (i)	v_i	w_i	v_i/w_i
1	20	5	4
2	12	4	3
3	10	5	2
4	8	4	2

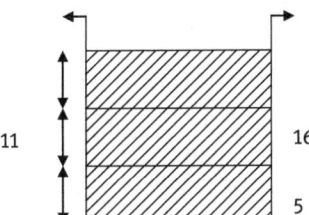

11

16

5

Figure 3.2: Solving the fractional knapsack problem with greedy approach.

Now, taking decision one by one.

For first item, $w_1 = 5 < U(=16)$ so

$x[1] = 1.0$

$U = U - w_1 = 16 - 5 = 11$

Again selection $w_2 = 4 < U(=11)$

$x[2] = 1.0]$

$U = U - w_2 = 11 - 4 = 7$

Selecting $w_3 = 5 < U(=7)$

$x[3] = 1.0$

$U = U - w_3 = 7 - 5 = 2$

Now for 4th item

$W_4 = 4, U = 2$

$W_4 > U$

So $x[4] = U/w4 = \frac{2}{4} = \frac{1}{2} = 0.5$

So solution vector is

$X[1:4] = \{1.0, 1.0, 1.0, 0.5\}$

Profit earnest is

$Z_{max} = v_1.x_1 + v_2.x_2 + v_3.x_3 + v_4.x_4$

$= 20*1 + 12*1 + 10*1 + 8*0.5 = 46$

Thus, the maximum profit earned is 46.

3.2 Backtracking

American mathematician D. H. Lehmer first proposed the term **backtrack** in the 1950s. The language SNOBOL (1962) was the first to provide a built-in general backtracking service. This approach is specially applied to NP-complete problems where no other choice of algorithm that may give better solution exists. The key point of backtracking algorithm is binary choice *Yes* or *No*, whenever backtracking has choice *No*, that means the algorithm has encountered a dead end. And it backtracks one step and tries a different path for choice *Yes*. The solution for the problem according to backtracking can be represented as implicit graph. In backtracking approach, we form a state space tree. This approach is used for a number of NP-hard problems.

3.2.1 Hamiltonian cycle

Hamiltonian cycle of a graph is a circuit that passes just once through each vertex of the graph G = (V, E) excluding the terminating vertex.

or

Hamiltonian cycle is a graph G round trip path in which each vertex is visited exactly once and returns to its starting point. Such graph is called Hamiltonian graph (see figure 3.3).

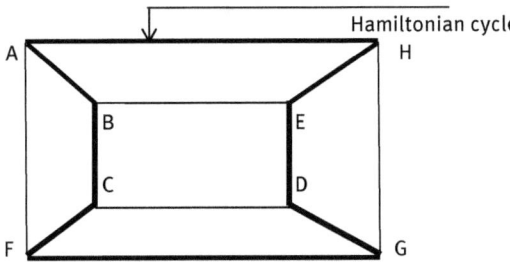

Hamiltonian cycle

Figure 3.3: Graph G with Hamiltonian cycle (in dark lines).

Without lack of generality, we can conclude that there is a Hamiltonian loop and it begins to form vertex "a" (Figure 3.4(a)) and we make vertex "a" root of the space tree in body (figure 3.4 (b)).

Let us use alphabetic order to break the tie between the vertices adjacent to a, we select the vertex b, then c, then d, and then e and finally to f which proves to be a dead end. So backtrack algorithm from f to e, and then to d and then c.

It offers the first alternatives for the algorithm to follow, moving c to e finally proves futile, so the algorithm has to backtrack from e to c, then to b. From there it goes to the vertices f, e, c, and d, from where the Hamiltonian circuit will regimentally return to yield: a, b, f, e, c, d, a (figure 3.4(c)).

Start of the Hamiltonian Cycle

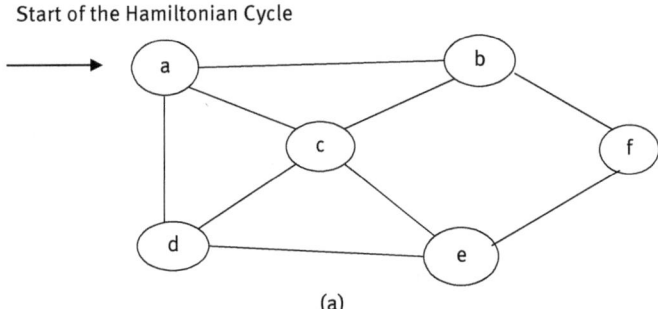

(a)

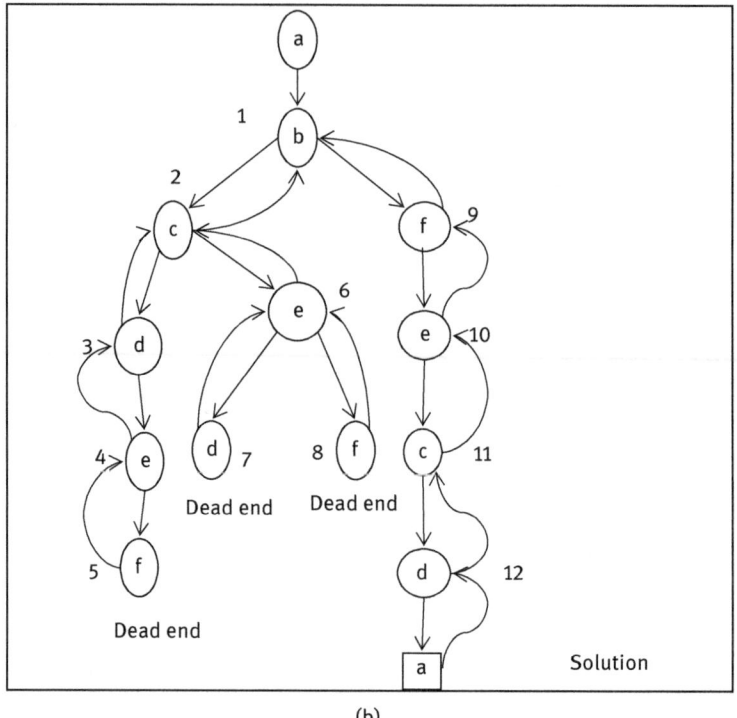

(b)

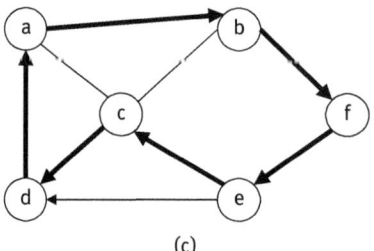

(c)

Figure 3.4: (a) A Graph, (b) Space Tree for Hamiltonian Cycle, (c) Hamiltonian Cycle.

Solution is a Hamiltonian cycle.

3.2.2 8-queen problem

The 8-queen problem is a problem in which 8 queens are to be placed on a 8 × 8 cross board so that no two queens are in attacking position. Same row, same column, and same diagonal are called the attacking positions. Backtracking algorithm is applicable to use wide range of problem and it is simple method. The backtracking resembles a depth first search tree in a directed graph.

Table 3.3: 8 × 8 board for 8-queen problem.

	1	2	3	4	5	6	7	8
1								
2								
3								
4								
5								
6								
7								
8								

Total block 8 × 8 = 64.

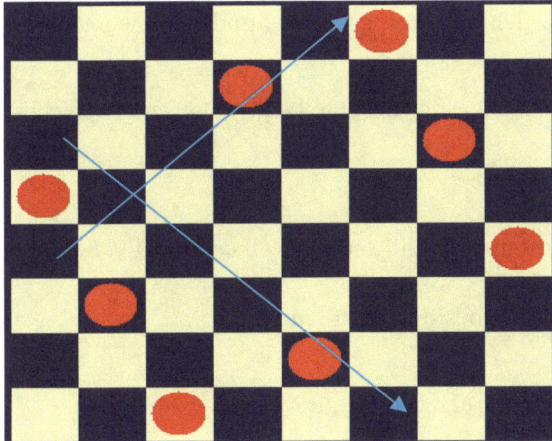

Figure 3.5: Possible solution for 8-queen problem.

The problem can be computationally quite expensive as there are C_8^{64}, that is, 4,426,165,368 possible placements of the queens on board, out of which only 92 arrangements satisfy the objective.

It is quite feasible to use shortcuts that reduce computational requirements or guidelines of thumb that avoids brute-force computational strategies. As an instance, just by way of making use of an easy rule that constrains every queen to an single column (or row), though nevertheless considered brute force, it is possible to lessen the quantity of possibilities to simply 16,777,216 (that is, 8^8) viable combos. Generating diversifications in addition reduces the opportunities to just 40,320 (this is, 8!), which are then checked for diagonal attacks. However, by allowing one placement of a queen on distinct row and column.

The number of ways are $\lfloor 8 = 40,320$ {8 Tuples}

Let us assure that ith queen is placed on row i so all the solution to 8-queen problem can therefore be represented by 8 tuples $(c_1, c_2,...,c_8)$ where c_i is the column on which ith queen is placed. The size of sample space is 8^8. Out of 8 tuples, no c_i's can be same, this reduces the size of sample space to $\lfloor 8$.

State space tree:

Backtracking algorithms decide problem solutions via systematically looking the answer space for the given problem instance. This seek is aided using a tree agency for the answer space. We start with an initial state node and organize a tree of solution space. Every node on this tree defines a problem state. All paths from the root node to other nodes define the state space of the problem. Solution states are those problem states for which the path from root to s defines a tuple within the solution space. The tree organization is known as state space tree.

Live node: A node is a live node which has been generated and all of whose child nodes have not yet been generated.

Dead node: This node is a generated node which is not to be more expanded on chess board.

(R-L) "UPPER LEFT TO LOWER RIGHT"

Again (R-L) value

$(5,1) = 6, (2,4) = 6, (3,3) = 6$ "lower left to upper right" (sum of cell coordinate). Similarly upper left to lower right $(3, 1) = 2, (4, 2) = 2$

Let us suppose two queens are placed at position (m,n) and (p,q)

Then they are on same diagonal

$$\text{if } m - n = p - q \text{ or } m + n = p + q$$

$$n - q = m - p \text{ or } n - q = p - m$$

Therefore, two queen lie on the same diagonal if

$$|n - q| = |m - p|$$

Algorithm: *Place (p, m)*
  //This returns a Boolean value that is true if m^{th} column is suitable for the pth queen.

```
{
    for n := 1 to (p-1) do
```

```
    if ( ( c [n] == m) // same column
    or [Abs (c[n] - m) == Abs (n-p)]) //diagonally attacking position
                 then return false;
  return true ;
}
```


Algorithm: *Nqueen (p, N)*

```
{
   for m: = 1 to N do
   {
     if place (p, m) then
     {
        c[p]: = m,
        If (p = N)
        Then write (c[1 : N]);
      else Nqueen (p+1, N);
      }
   }
}
```

All solution of the N-queens problem for n = 8 are 92.One of the possible solution of 8-queen problem is:

$$\{c_1, c_2, c_3, c_4, c_5, c_6, c_7, c_8\} \equiv \{4,6,7,8,2,7,1,3,5\}$$

3.3 Dynamic programming

Dynamic programming is an innovative way for fixing complex problems by means of breaking them into some small or tiny subproblems or subtasks. It is applicable for issues exhibiting the residences of overlapping subproblems and top-quality substructure (described below). When applied, the method takes far less time than naive methods that do not take gain of the subproblem overlap (like depth-first search).

Dynamic programming is a method for fixing issues with overlapping subproblems. These subissues rise up from a recurrence relating option to a given hassle with solution to its smaller subproblems of the identical type. Rather than fixing overlapping subissues repeatedly. Dynamic programming advises solving every smaller subproblem just once and storing the result in a table from which we can obtain a solution to the original problem.

The Fibonacci relation: $F_n = F_{n-1} + F_{n-2}$

$$F_0 = F_1 = 1 \text{ (initial conditions)}$$

A recurrence relation, if we know F_{n-1} and F_{n-2}, we can compute F_n easily.

3.3.1 0/1 knapsack problem

We have n items of weights w_1, w_2, ...w_n and values v_1, v_2, ... v_n, and W is the capacity of knapsack. We need to fill the knapsack while gaining maximum profit. In 0/1 knapsack, we cannot take portion of object, that is, either an object is selected for knapsack or not.

The 0/1 knapsack problem as LPP is:

$$\text{Max } z = \sum_{i=1}^{n} x_i.v_i$$

$$\text{such that} \sum_{i=1}^{n} x_i.w_i \leq W$$

$$v_i \geq 0, w_i \geq 0 \text{ and } x_i \in \{0, 1\}$$

We are to find a recurrence relation that expresses a strategy to an example in phrases of its small subinstances. The time and space performance each for the set of rules is $\theta(nW)$.

Let us take the first i items $(1 \leq i \leq n)$, and knapsack ability $j < W$. $O[i, j]$ be the value of optimal solution of these i items that suit the knapsack of capability j. This instance of i items is partitioned into subsets of two categories,
- First subset does not contain the ith item, $O[i-1, j]$ is the value of this subset. If ith object is not placed into the knapsack then the value will be same as of i−1 objects, so: $O[i, j] = O[i-1\ j]$
- Second subset contains the ith item, therefore the value of such optimal subset is

$$v_i + O[i-1, j-w_i]$$

Thus, the value of an optimal solution of the first i objects is the maximum of these two categories' values.

$$O[i, j] = \max\{ O[i-1, j],\ vi + O[i-1, j-w_i]\ \text{if}(j-w_i) \geq 0$$

The initial conditions are:

$$O[0, j] = 0 \text{ for } j \geq 0 \text{ And } O[i, 0] = 0 \text{ for } i \geq 0$$

The objective is to find $O[n, W]$.

Example 3.3: We have the following information

Item	Weight	Value
1	2	RS 12
2	1	RS 10
3	3	RS 20
4	2	RS 15

CAPACITY
W = 5

Item/Capacity	0	1	2	3	4	5
0	0	0	0	0	0	0
1	0	0	12	12	12	12
2	0	10	12	22	22	22
3	0	10	12	22	30	32
4	0	10	15	25	30	37

Figure 3.6: Knapsack problem.

$$O[0, J] = 0, J \geq 0 \text{ and } O[i, 0] = 0, i \geq 0$$

$O[4,5] > O[3,5]$, item 4 is included in knapsack . Again $O[3,3] = O[2, 3]$, so item 3 is not included in knapsack. Also $O[2,3] > O[1, 3]$, so item 2 is included in knapsack. Again $O[1, 2] > O[0, 2]$, thus item 1 is also included. Hence optimal solution is {1, 2, 4}. (see figure 3.6).

Optimal value is $O[4, 5]$ = RS. 37

3.3.2 The traveling salesman problem

We have a weighted, undirected, and complete graph $G = (V, E)$. Each edge (u, v) has positive integer cost $c(u,v)$. The travelling salesman problem is to start from source vertex and cover all the vertices exactly and then return back to starting vertex and the cost of route must be minimum, that is, we have to find out Hamiltonian cycle of G with minimal cost. The cost $c_{ij} = \infty$ if $(i,j) \notin E$, $|V| = n$ where $n > 1$.

Let $g(i, S)$ be the length of shortest path beginning at vertex i going via all vertices in S and ending at vertex 1.The value $g(1, V\text{-}\{1\}$ is the length of optional tour, so we have

$$g(1, V - \{1\}) = \min_{2 \leq k \leq n} \{c_{1, k} + g(k, V - \{1, k\})\}$$

We have to obtain,

$$g(i, S) = \min_{J \in S} \{C_{ij} + g(j, s - \{j\})$$

Clearly, $g(i, \phi) = c_{i1}, 1 \le i \le n$

When $|S| < n - 1$

The values of i and S for which g(i, S) is needed are such that $i \ne 1$, $1 \notin S$, $i \notin S$.

$$g\{1, V - \{1\}\} = \min_{2 \le k \le n} \{c_{1k} + g\{k, V - \{1, k\}$$

$$g\{i, s\} = \min_{j \in s} \{c_{ij} + g\{j, (S - j)\}$$

Clearly $g(i, \phi) = c_{i1}$

Example 3.4: Solve the given TSP in Figure 3.7

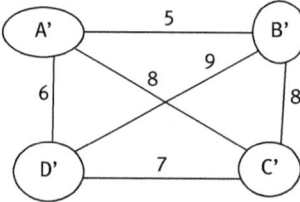

Figure 3.7: The traveling salesman problem.

Since $g\{i, \phi\} = c_{i1}$ so

$$g(B', \Phi) = c_{B' A'} = 5$$

$$g(C', \Phi) = c_{C' A'} = 8$$

$$g(D', \Phi) = c_{D' A'} = 6$$

Now compute g(i, S) with $|S| = 1$,

$$g(B', \{C'\}) = \{c_{B' C'} + g(C', \phi)\} = 8 + 8 = 16$$

$$g(C', \{D'\}) = \{c_{C' D'} + g(C', \phi)\} = 13$$

$$g(D', \{C'\}) = \{c_{D' C'} + g(C', \phi)\} = 15$$

$$g(C', \{B'\}) = \{c_{C' B'} + g(B', \phi)\} = 13$$

$$g(B', \{D'\}) = \{c_{B' D'} + g(D', \phi)\} = 9 + 6 = 15$$

$$g(D', \{B'\}) = \{c_{D' B'} + g(B', \phi)\} = 9 + 5 = 14$$

When |S| = 2,

$$g(B', \{C', D'\}) = \min[\{c_{B'C'} + g(C', \{D'\}), \{c_{B'D'} + g(D', \{C'\}]$$
$$= \min[(8 + 13), (9 + 15)]$$
$$= \min[21, 24]$$
$$= 21$$

$$g(C', \{B', D'\}) = \min[\{c_{C'B'} + g(B', \{D'\}), \{c_{C'D'} + g(D', \{B'\}] = 21$$

$$g(D', \{B', C'\}) = \min[\{c_{D'B'} + g(B', \{C'\}), \{c_{D'C'} + g(C', \{B'\}] = 20$$

Finally,

$$g(A', \{B', C', D'\}) = \min[\{c_{A'B'} + g(B', (C', D'))\}, \{c_{A'B'} + g(C', \{B', D'\})\},$$
$$\{c_{A'D'} + g(D', \{B', C'\})\}]$$
$$= \min\{26, 29, 26\} = 26$$

So, the length of shortest tour is 26 and the optimal tour is A'-D'-C'-B'-A'.

3.3.3 Chain matrix multiplication

The chain matrix multiplication problem is important in the area of compiler design and database. As we want to code optimization and query optimization. The problem statement is:

Given {A_1, A_2 . . . . .A_n}, matrix A_i has dimension p_{i-1} x p_i and we have to perform the matrix multiplication and parenthesize the product A_1 X A_2 . . . . A_n in such a way so that the number of scalar multiplications will be minimum.

Since matrix multiplication holds associativity property but not commutativity, we have to follow above matrix order for multiplication but we are free to parenthesize the above multiplication. Two matrices are compatible for multiplication if number of columns in first matrix equals to rows in second.

$$C[i][j] = \sum_{k=1}^{q} A[i][k] * B[k][j]$$

The number of places in parenthesis in sequence of n matrices given by p(n). Then,

$$p(n) = 1 \text{ for } n = 1$$

Let us insert split between matrices, subproduct may take place between mth and m + 1th matrices. Thus, for any m = 1 to n−1, the recurrence relation will be,

$$p(n) = \begin{cases} 1, & \text{if } n = 1 \\ \sum_{m=1}^{n-1} p(m).p(n-m) & \text{if } n \geq 2 \end{cases}$$

Let $A_{i \ldots j}$ represent the product of A_i, $A_{i+1} \ldots \ldots \ldots A_j$, where $i \leq j$. If problem is nontrial, $i > k$ then any parenthesization of the product A_i, $A_2 \ldots \ldots \ldots A_j$ must split the product between A_k and A_{k+1} for some integer k where $i <= k < j$. For some value of k, we first compute $A_{i \ldots k}$ and $A_{k+1 \ldots j}$ and then result is combined to get the final product $A_{i \ldots .j}$. The cost of this process is therefore the sum of computing cost of matrix $A_{i \ldots k}$ and $A_{k+1 \ldots j}$ and cost of multiplying these two matrices together. To compute the matrix $A_{i \ldots j}$, suppose minimum N[i, j] number of scalar multiplication are performed. N[i, j] can be defined as:

$$N[i, j] = N[i, k] + N[(k+1)j] + \{ p_{i-1} x p_k x p_j \}, \text{where } i \leq k < j \text{ Clearly, } N[i, i] = = 0.$$

The orders of A_i is P_{i-1} x P_i, so the order of $A_{i \ldots .j}$ is P_{i-1} x P_j. We assume that the value of k is already known, and there can be (j–i) possible values of k. So, the recursive equation which gives minimum value is,

$$N[i, j] = \begin{cases} 0, & \text{if } i = j \\ \min_{i \leq k < j} \{ [N[i, k] + N[(k+1), j] + (p_{i-1} x p_k \times p_j) \} & \text{if } i < j \end{cases}$$

p_{i-1} x p_k × p_j gives number of scalar multiplications required for the product of $A_{i \ldots K}$ and $A_{k+1 \ldots j}$.

3.3.3.1 Optimal cost algorithm

Order of matrix A_i is p_{i-1} x p_i, and the input is a sequence $P = < p_0, p_1 \ldots \ldots p_n >$, where length of the sequence is $n + 1$.

Algorithm: *Matrix-Chain-Order (P)*
   // Table S[1..n,1..n]is used to store at which index of k the optimal cost is achieved in computing N[i .j].

```
{
      L : = length (P)-1;
      for j := 1 to L do
             N[j, j] : = 0; // Minimum cost for chain length 1//
      for cl:= 2 to L do // cl is chain length
             {
                   for j := 1 to (L-cl+1) do
                          {
                                i := (j + cl-1);
                                N[j, i]: = ∞; // Intial Value
                                for k : = j to (i-1) do
                                       {
                                             c : = N[j, k] + N[(k+1) i] + p_{j-1}. p_k. p_i;
                                             if c< N[j, i]
                                                    N[j, i]: = c;
```

                                                          S[j, i]: = k;
                                            }
                                }
                    }
                    return N and S.
}

The following recursive procedure computes the matrix chain product, the table S computed by previous algorithm and the indices i and j. The initial call is Print-Optimal-Patterns (A, S, 1, n).

Algorithm: *Print-Optimal-Patterns(A, S, p,q)*

```
{
    if (p==q)
Then print "A_p"
    else print "("
        Print-Optimal-Patterns (A, S, p, S[p, q]);
        Print-Optimal-Patterns (A, S, S[p, q]+1, q);
        print ")"
}
```

3.3.4 Optimal binary search tree

A binary search tree (BST) is a kind of binary tree which satisfies the following properties:
1) All the elements have distinct key values.
2) The keys (if exist) in left subtree are smaller than the key at the root.
3) The keys (if exist) in right subtree are large than the keys at the root.
4) The left and right subtrees are also binary search tree (see figure 3.8).

The total number of BST with n keys is equal to the n-Catalon number,

$$C(n) = C_n^{2n}(\frac{1}{n+1})$$

Let $a_1, a_2, \ldots \ldots a_n$ be the distinct keys ordered from the smallest to largest, and let $p_1, p_2, \ldots \ldots \ldots p_n$ be the probability of searching them. Let c[i,j] be the smallest average number of comparison made for successful search in binary tree T_i^j. Made of keys $a_i, a_{i+1} \ldots .a_j$ where i and j are integer indices and $1 \le i \le j \le n$. We will find values of c[i,j] for all smaller instances of the problem although we are interested just in c[1, n].

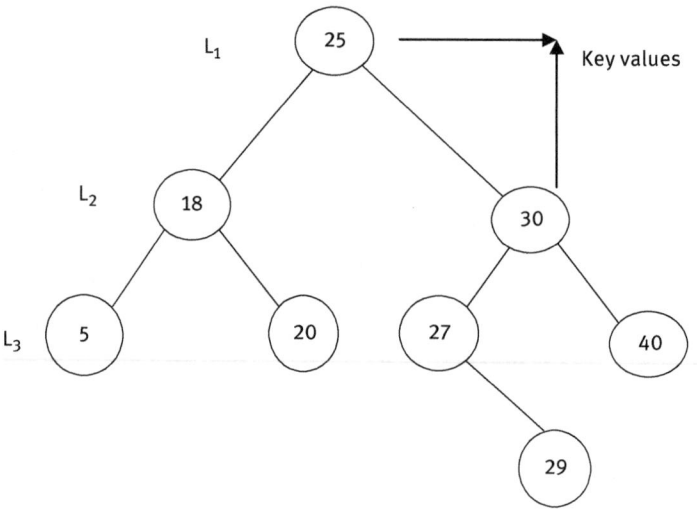

Figure 3.8: A binary search tree.

We need to consider all possible ways to choose a root a_k among the keys a_i, a_{i+1}, ...a_j for such a binary tree, the root contains key a_k. The left subtree T_i^{k-1}contains keys a_i, ...a_{k-1} optimally arranged and the right subtree T_{k+1}^j contains keys $a_{k+1, -,}$ a_j also optimally arranged (see figure 3.9). If we count tree level starting with 1 then we have,

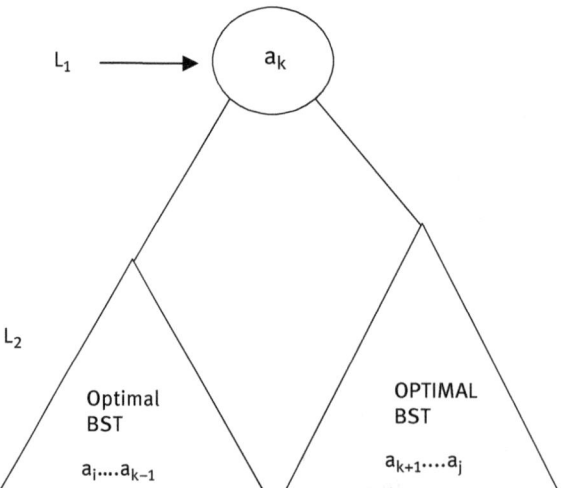

Figure 3.9: A binary tree.

$$\text{Mathematical expression } E(x) = \sum_{i=1}^{n} p_i x_i$$

$$c[i,j] = \min_{i \leq k \leq j} \left\{ P_k.1 + \sum_{s=i}^{k-1} p_s.\{\text{level of } a_s \text{ in } T_i^{k-1} + 1\} + \sum_{s=k+1}^{j} p_s.\{\text{level of } a_s \text{ in } T_{k+1}^{j} + 1\} \right\}$$

$$= \min_{i \leq k \leq j} \left[\left\{ \sum_{s=i}^{k-1} p_s.\{\text{level of } a_s \text{ in } T_i^{k-1}\} + \sum_{s=k+1}^{j} p_s.\{\text{level of } a_s \text{ in } T_{k+1}^{j}\} + \sum_{s=i}^{j} p_s \right] \right.$$

$$c[i,j] = \min_{i \leq k \leq j} \{ c[i, k-1] + c[k+1, j] \} + \sum_{s=i}^{j} p_s, \text{ where } 1 \leq i \leq j \leq n$$

In empty tree, $c[i, i-1] = 0$ for $1 \leq i \leq n+1$.

Number of comparison in a tree containing one element $c[i, i] = p_i$ for all $1 \leq i \leq n$.

A two-dimensional (2-D) table shows volumes needed for computing $c[i, j]$ (see table 3.4).

Table 3.4: 2-D table.

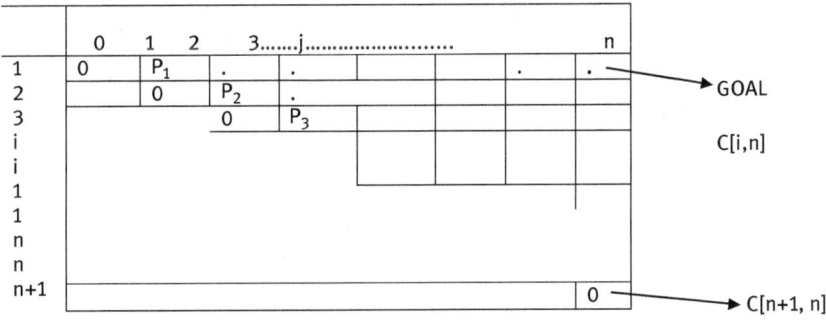

In the table, entries in the main diagonal all are zero, and given probabilities $p_i, 0 \leq p_i \leq 1$ right above it and moving towards the upper right corner. The final computation is for $c[1,n]$ gives the average number of comparison for successful search in optimal binary tree. Also, to get optimal tree, we need to maintain another 2-D table to record the values of k for which is key comparison minimum, is achieved.

3.4 Branch-and-bound technique

The branch-and-bound technique like backtracking explores the implicit graph and deals with the optimal solution to a given problem. In this technique, at each state, the bound is calculated for a particular node and is checked whether this bound will be able to give the solution or not. That means we calculate how far we are from the solution in the graph. At each stage, there is a lower bound for minimization problems and an upper bound for maximization problems. For each node, bound is calculated by means of partial solution (PS). The calculated bound for the node is checked with previous best result and if found that new PS results lead to worse case, then bound with the best solution so far, is selected and we leave this part without exploring it further. Otherwise, the checking is done with the previous best result obtained so far for every PS while exploring.

In branch-and-bound technique, we need a function whose value is computed for each node of state space tree. This function helps in identifying the promising and nonpromising nodes in state space tree and hence helps in reducing the size of the tree. Also, it accelerates the algorithm. A good function is not easy to find and this function also should be easily computable.

3.4.1 Assignment problem

In assignment problem, "n" facilities are assigned "n" tasks. We have to find the minimum total cost of assignment where each facility has exactly one task to perform. We are given a n x n cost matrix C, where C_{ij} refers to the minimum cost for person P_i to perform task T_j, $1 \le i \le n$ and $1 \le j \le n$, then the problem is to assign facility to task so that the cost will be minimum.

Example 3.5:
Job

$$
\text{Facilities} \quad
\begin{array}{c}
\\ A \\ B \\ C \\ D
\end{array}
\begin{array}{c}
\text{Job} \\[4pt]
\left[
\begin{array}{cccc}
j_1 & j_2 & j_3 & j_4 \\
9 & \boxed{2} & 7 & 8 \\
6 & 4 & \boxed{3} & 7 \\
5 & 8 & \boxed{1} & 8 \\
7 & 6 & 9 & \boxed{4}
\end{array}
\right]
\end{array}
\quad 4 \times 4
$$

It is obvious that the cost of any solution, including optimal, cannot be smaller than the sum of the smallest elements in each row of matrix, lower bound = 2 + 3 + 1 + 4 = 10

At each level of the state space tree (see figure 3.10), the node with the minimum cost is selected for exploring. Root node signifies that no assignment is yet done. At next level, the assignment a→2 corresponds to minimum cost so selected for exploring and so on. The best solution a →2, b→1, c → 3 d →4. And the cost will be 13.

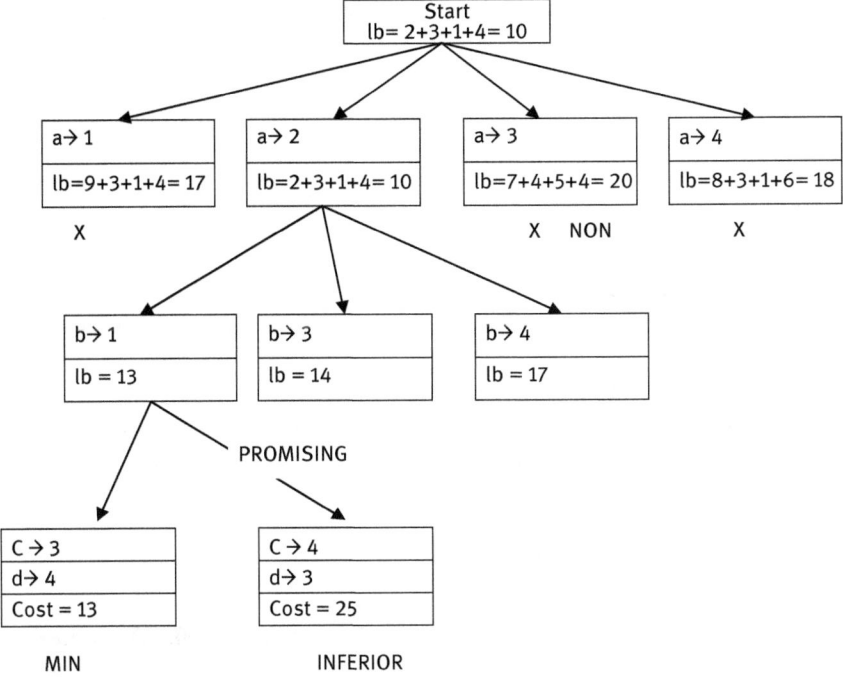

Figure 3.10: State space tree of example 3.5.

The minimum cost =13, optimal assignment is a–>2, b–1, c–>3, d–>4.

3.4.2 Knapsack problem

We have to place n objects in a given knapsack. The knapsack capacity is W and v_i is the value earned by placing ith item. The aim is to find the most valuable subset of items that fit in the knapsack. The objects are sorted in descending order by the value to weight ratios, that is,

$$\frac{v_1}{w_1} \geq \frac{v_2}{w_2} \geq \ldots \ldots \ldots \frac{v_n}{w_n}$$

The simple way to compute the upper bound (ub) is to add v of the item already selected, the product of the remaining capacity of the knapsack, (W-w) and the best per unit pay of + amount remaining items which

$$u_b = v + (W - w)(V_{i+1}/w_{i+1})$$

Example 3.6: Solve knapsack problem given the item list as follows (in table 3.5)

Table 3.5: Knapsack problem.

Item	Weight	Value	Value/weight
1	4	40	10
2	7	42	6
3	5	25	5
4	3	12	4

The knapsack capacity (W) is 10.

The root of the state space tree, no items have been selected yet as both total weight of the items are already selected w and total value v are equal to 0.

The upper bound ub = 0 + 10 x 10 = 100

$$ub = v + (W - w)\left(\frac{v_{i+1}}{w_{i+1}}\right)$$

Best solution of knapsack by branch and bound is subset of items {1,3}, Max value = 65 (see figure 3.11).

3.5 Amortized analysis

In amortized analysis, the time required to perform a sequence of data structure operations is average of all operation performed. It is different from average case analysis since it such analysis probability doesn't involves. We computer an upper bound $T(n)$ on total cost of a sequence of n operations. Then the average cost per operation is $T(n)/n$. This average cost is taken as amortized cost so that all operations have the same amortized cost.

3.5.1 Aggregate method

In aggregate analysis, a series of n operations consumes worst-case time $T(n)$ in total. In worst case, the average case or amortized cost per operation is $T(n)/n$.

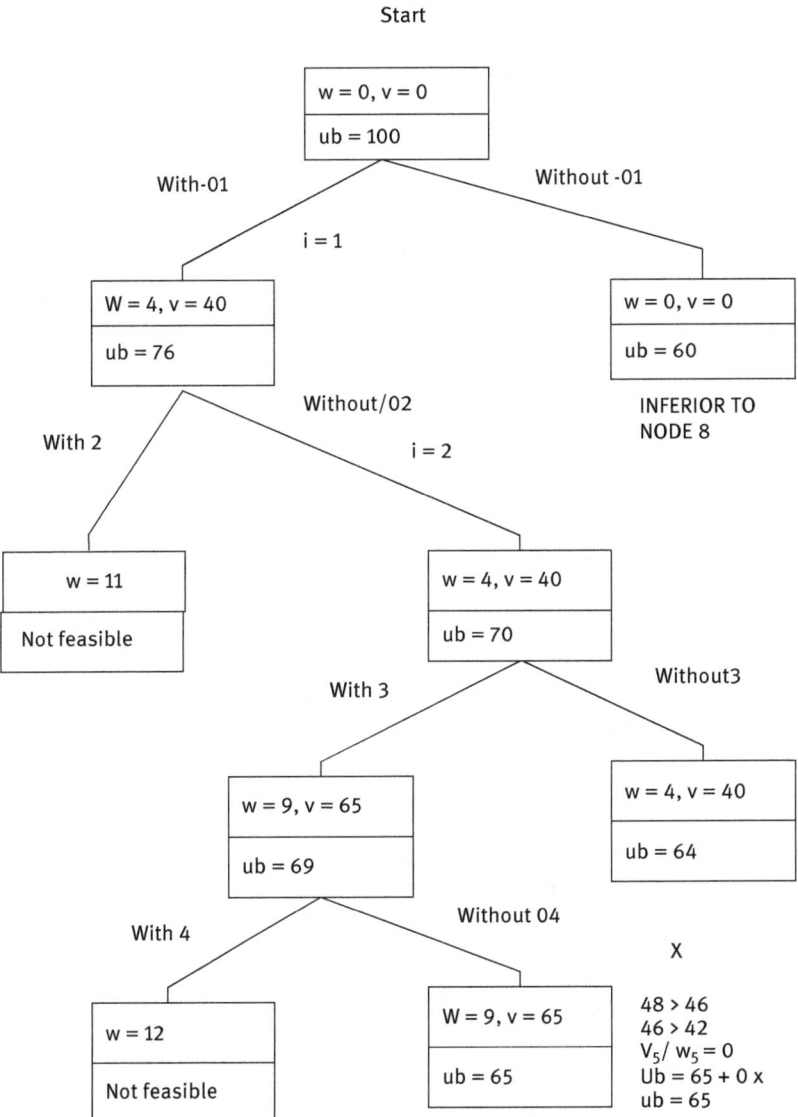

Figure 3.11: Solution space for knapsack problem.

3.5.2 The potential method

The potential method stands for prepaid efforts as potential that can be released to pay for future operations. Assume that DS_0 is the starting data structure on which n operations are carried out and the result of applying the ith operation is the data structure DS_i and c_i be the actual cost of the ith operation. A potential function maps each DS_i to a real number $\phi(DS_i)$ which is the potential related with that data structure.

The amortized cost $\hat{c}_i$ of i_{th} operation can be defined as,

$$\hat{c}_i = c_i + \Delta\phi_i$$

$$\Delta\phi_i = \phi(DS_i) - \phi(DS_{i-1})$$

So the amortized cost of each operation is sum of actual cost and the increase in potential. The total amortized cost will be (for n operations),

$$\sum_{i=1}^{n} \hat{c}_i = \sum_{i=1}^{n} (c_i + \phi(DS_i) - \phi(DS_{i-1}))$$

$$= \sum_{i=1}^{n} c_i + \phi(DS_n) - \phi(DS_0))$$

if $\phi(DS_n) \geq \phi(DS_0)$ Then, total amortized

cost $\sum_{i=1}^{n} \hat{c}_i$ is an upper bound on the total actual cost $\sum_{i=1}^{n} c_i$

If $\phi(DS_i) - \phi(DS_{i-1}) > 0$, then the amortized cost $\hat{c}_i$ represents an overcharge to the ith operation and the potential of data structure increases. If potential difference is negative then the amortized cost $\hat{C}_i*$ represents an undercharge of the ith operation and actual cost of the operation is paid by the decrease in the potential.

Example 3.7: There are three stack operations, PUSH, POP, and MULTIPOP. PUSH simply places an element on the top of the stack and POP operation deletes an element from the stack. MULTIPOP operation deleted multiple elements at once. Define a potential function ϕ on the stack to be the number object in the stack. For empty stack

$\phi(DS_0) = 0$ $\qquad$ $\phi(DS_i) \geq 0$ The number of object in stock is never negative.
Since $\phi(DS_0) = 0$
$\Rightarrow \phi(DS_i) \geq \phi(DS_0)$

Thus, the total amortized cost of n operation with respect to ϕ represents upper bound on the actual cost. If the ith operation on the stack is PUSH, while stack contains p objects then the potential difference will be,

$$\phi(DS_i) - \phi(DS_{i-1})) = p + 1 - p = 1$$

The amortized cost of PUSH operation is

$$\hat{c}_i = c_i + \phi(DS_i) - \phi(DS_{i-1})$$

$$\hat{c}_i = 1 + 1 = 2, 2 \in O(1)$$

Suppose the ith operation is MULTIPOP(p, k) and $k' = min(p, k)$ objects are popped off the stack. The actual cost of the operation is k' potential difference,

$$\phi(DS_i) - \phi(DS_{i-1}) = -k'$$

Thus the amortized cost of MULTIPOP,

$$\hat{c}_i = c_i + \phi(DS_i) - \phi(DS_{i-1})$$

$$\hat{c}_i = k' - k' = 0$$

Similarly, the amortized cost of an ordinary POP operation is 0. The amortized cost of each operation is O(1). So total amortized cost of n operations will be O(n).

3.6 Order statistics

Order set Š: The set in which elements are arranged in ascending order,

$$Š = \{2, 3, 4, 5, 6\}$$

Let Š be an ordered set containing n elements, if n is odd, then (n + 1) is even.

Then, median is given by $\left\lceil (\dfrac{n+1}{2}) \right\rceil$

Again, if n is even, then median are given by $n/2$ and $n/2 + 1$,

$$\text{or} \left\lfloor \dfrac{n+1}{2}) \right\rfloor \text{ and } \left\lceil -\dfrac{n+1}{2}) \right\rceil$$

Lower median Upper median

$$\text{let } Š = \{18, 13, 12, 14, 10\}$$

$$\text{order set } Š = \{10, 12, 13, 14, 18\}, n = 5$$

$$\text{median} = \left\lceil \dfrac{n+1}{2}) \right\rceil = \left\lceil \dfrac{5+1}{2}) \right\rceil = \lceil 3 \rceil = 3$$

$$\text{Again, if } Š = \{1, 2, 3, 4, 5, 6\} n = 6$$

$$\text{Median are } \left\lfloor \dfrac{n+1}{2}) \right\rfloor = \lceil 3.5 \rceil = 3, \text{ and } \left\lceil \dfrac{n+1}{2}) \right\rceil = \lceil 3.5 \rceil = 4$$

i^{th} order statistics

If Š is an ordered set containing n elements, then ith order statistics for Š is the ith smallest element of S.

If i = 1, the ith order statistics gives smallest element of the set and for i = n, we get max element of the set.

Algorithm: *Median (S)*
{ // S is an ordered set containing "n" elements,

Set n: = length (S);

if((n/2) = = 0) hen

$$i := \left\lfloor \frac{n+1}{2}) \right\rfloor, \quad j := \left\lceil \frac{n+1}{2} \right\rceil;$$

returni, j;

else

$$k := \left\lceil \frac{n+1}{2}) \right\rceil;$$

return k;

}

This algorithm completes in a constant time because every steps takes a constant amount of time (O(1)). So, the complexity will be $\theta(1)$.

Algorithm: *Selection (i)*
{ // Selection of ith smallest element of set S, first changing S into order set.

S: = sort (S);

p := s[i]

return p; // p is the ith smallest element of set Š.

The complexity of this algorithm is based on the complexity of the sorting algorithm if sorting algorithm is heap sort or merge sort. The complexity of algorithm is $O(n\log_2 n)$.

Algorithm: *Min(S)*
{ // finding the minimum element of a given set S containing p element.

minimum : = s[1] // setting first element as min element

for j:= 2 to p do

{

if(minimum>s[j]) Then

minimum: = s[j];

}

Return (minimum);

}

The complexity of this algorithm depends on the basic operation comparisons, total comparison are $(n-1)$ so complexity is $\theta(n)$.

Problem set

1. Explain the concept of dynamic programming with an example.
2. Compare and contrast dynamic programming with divide-and-conquer (DAC) programming?
3. Explain in detail the primary difference between the two techniques: DAC and dynamic programming?
4. What are the fundamental features of dynamic programming?
5. Calculate the value of C(12,8) using dynamic programming.
6. Solve the fractional knapsack problem with the greedy algorithm. The knapsack capacity W = 60.

Item (i)	Weight (Wi)	Profit Vi
1	5	30
2	10	20
3	20	100
4	30	90
5	40	60

7. Explain the applications of backtracking.
8. Compare and contrast brute force approach and backtracking.
9. Write the control abstraction of backtracking.
10. Use Branch and Bound technique to solve the assignment problem:

	F_1	F_2	F_3	F_4
J_1	4	3	1	5
J_2	6	2	9	8
J_3	3	4	7	6
J_4	2	8	4	9

Chapter 4
Advanced graph algorithm

This chapter introduces graph concept and algorithms based on graph.

4.1 Introduction

A graph is a collection of vertices and edges. These vertices are connected by edges. A graph can be defined by two sets V and E, where V is the set of vertices of the graph and E is the set edges. Figure 4.1 shows two graphs: each one has five vertices, graph A has six edges, and graph B has eight edges.

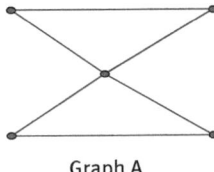
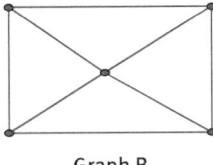

Graph A   Graph B

Figure 4.1: Graph A (a) and graph B (b).

4.1.1 Terminology

Here are some terminologies used throughout this chapter:
(i) **Subgraph:** As the name indicates it is also a graph. The *subgraph H* of a graph can be denoted as H = (V', E') such that V' $\in$ V and E' $\in$ E.
(ii) **Spanning Subgraph:** The spanning subgraph has all vertices of G but not necessarily edges.
(iii) **Connected Graph:** In connected graph, any of the two vertices have a path joining them.
(iv) **Adjacent Vertices:** Adjacent vertices are the end points of the same edge.
(v) **Neighbor Vertices:** If two vertices are adjacent to one another, they are said to be neighbor vertices.
(vi) **Isolated Vertex:** If a vertex has no neighbor vertex it is said to be an isolated vertex.
(vii) **Complete Graph:** If every vertex of the n vertices of graph G is adjacent to the other n−1 vertices of G (see Figure 4.2).

https://doi.org/10.1515/9783110693607-004

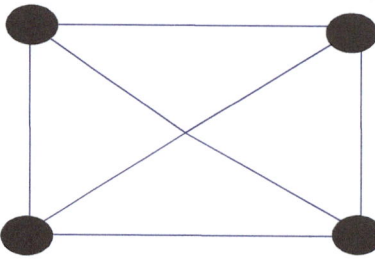

Figure 4.2: Complete graph.

(viii) **Tree:** A tree is a graph that is connected and has no cycles (see Figure 4.3).

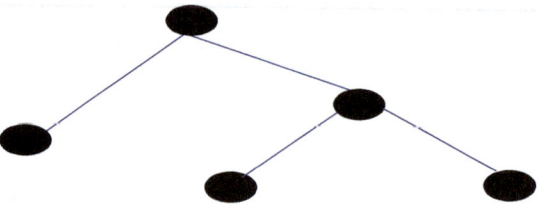

Figure 4.3: Tree.

4.2 The graph search techniques

There are two techniques for performing search operation on graphs, namely breadth first search (BFS) and depth first search (DFS).

4.2.1 Breadth first search

BFS starts with a vertex q and makes it as visited. This vertex q is currently said to be unexplored. Exploring this vertex means all adjacent vertices of this vertex are visited. Next step is to visit all adjacent vertices of vertex q and which are still unexplored

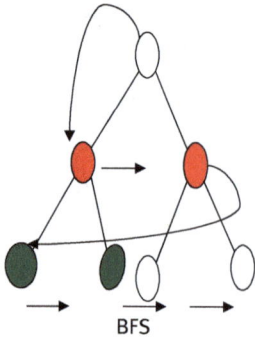

BFS

Figure 4.4: Breadth first search.

(see Figure 4.4). When all adjacent vertices of q are visited then vertex q is said to be explored. Unexplored vertices are maintained in a list. All these vertices are then added to this list. Next vertex to be explored is the first node in this list. This process of vertex exploration keeps on going until the list of unexplored is not empty.

Algorithm: *Breadth-First-Search (G)*
// Input: A Graph
// Output: All the vertices are marked with the integers depicting the order they have been visited.
Integer 0 indicates vertex p is not yet visited.

```
{    Mark each vertex in G with 0. // Not visited
     Visit_Order: =0; //global variable
     For all vertices p in V do
          If p has number 0 // not yet visited
          EXPLORE (p);
}
```

Algorithm: *EXPLORE (p)*

```
{
     Visit_Order := Visit_Order+1;
     Mark p with the value Visit_Order and initialize queue with p;
     While queue is not empty do
          For every vertex q in V adjacent to vertex p do
          If q in not yet visited // marked with 0
               Visit_Order := Visit_Order +1;
               Mark q with Visit_Order;
          Add q to the queue;
          Remove vertex p from the front of queue;
}
```

Complexity of BFS: Queue space need is at most (n–1), if q is connected with (n–1) remaining vertices, and then adjacent vertices of q are on the queue all at once therefore,

$$S(n, e) = \theta(n)$$

If Adjacency list is used for representation of graph G, then all adjacent vertices of p can be determined in time equivalent to the degree of p, that is, d(p). If G is directed graph, d(p) is the out degree of node p. The time for *for* loop in algorithm is of $\theta(d(p))$ since each vertex in G can be explored exactly once.

Example 4.1: Adjacency list representation Figure 4.5,

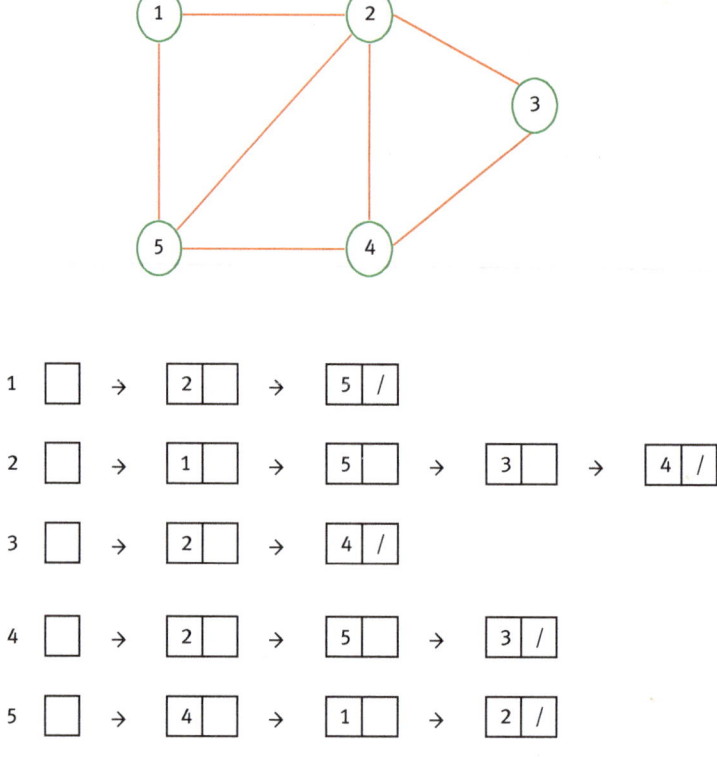

Figure 4.5: Graph and its adjacency list representation.

In *Breadth-First-Search(G)* algorithm, each vertex is enqueued and dequeued exactly once and operation of enqueue and dequeue takes O(1) time. Therefore total time for performing the operation enqueue is O(|V|) because the list is searched at most once, for dequeue vertex. The time spent in adjacency list scanning process is O(|E|). The initialization takes time O(|V|). Hence complexity of the algorithm is O(|V| +|E|), when the graph is represented by adjacency linked list. For adjacency matrix representation, it is O(|V|²).

4.2.2 Depth first search

DFS starts with a vertex p in G, p is said to be current vertex. Any edge (p, q) incident to the current vertex p is traversed by the algorithm. If this edge vertex q is already visited, then search backtrack to the vertex p. Otherwise go to q and q becomes new current vertex. This process is repeated until "deadend" is reached. Then back-tracking is performed. The process ends when backtracking leads back to the starting vertex. DFS is a recursive algorithm (figure 4.6).

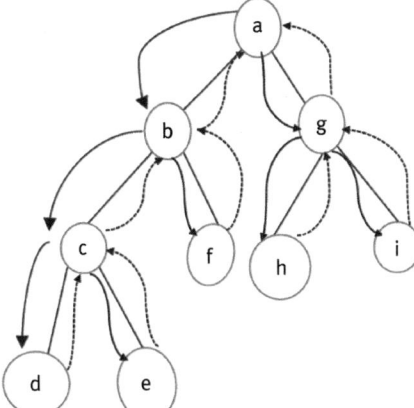

Figure 4.6: Depth first search.

Algorithm: *Depth-First-Search (G)*
// Input: A graph G
// Output: *All the vertices are marked with the integers depicting the order they have been vis-ited. Integer 0 indicates vertex p is unvisited.*

```
{     Mark each vertex in G with 0.
      Visit_Order: = 0;
      For each vetex p in V do
            If p is not yet visited
            DFS_VISIT (p);
}
```

Algorithm: *DFS_VISIT (p)*

```
{
      Visit_Order: = Visit_Order +1;          //global variable Visit_Order
      Mark p with Visit_Order.
      For each vertex q in V adjacent to p do
      If q is not yet visited
            DFS_VISIT (q);
}
```

Complexity of DFS: The complexity of algorithm will be O ($|V|^2$) if graph is represented by adjacency matrix. If the graph is represented by adjacency linked list the complexity will be O ($|V|+|E|$).

Example 4.2: DFS traversal of the graph (figure 4.7) given later.

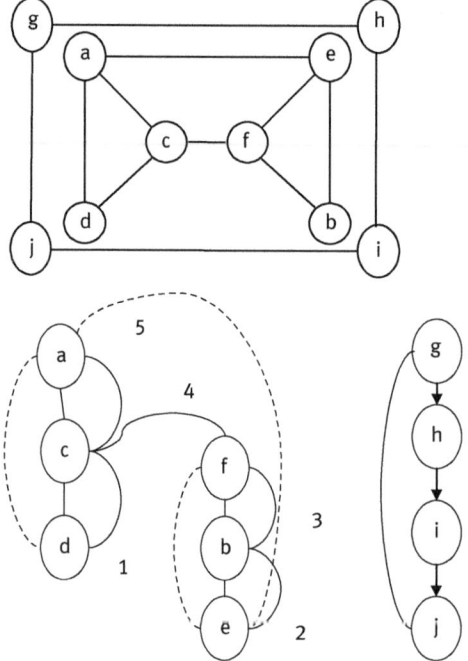

Figure 4.7: Graph and its DFS traversal.

4.3 Spanning tree

A spanning tree of an undirected and connected graph is a subgraph that is a tree and contain all the vertices of G. There can be more than one spanning trees of a graph. Weights can also be assigned to edges of the graph; this weight represents edge cost or length. Such graphs are called **weighted graph**. A **minimum spanning tree (MST)** is the spanning tree with minimum possible weight.

4.3.1 Kruskal's algorithm

This algorithm is used to find MST using greedy approach.

Algorithm: *Kruskals-MST (G)*
// Input: A connected weighted graph G (V, E)
// Output: E_{MST}: The set of edges creating a MST.
//Sort E: Arrange edge weights in increasing order
//w(e_{i1}) ≤ . . . . ≤ w($e_{i \ |E|}$),   *|E| Number of edges in G.*

```
{
     E_MST:≡ Null( Φ );
     counter: = 0;
     j: =0;
     While (counter < (|v|-1)) //TOTAL (|v|-1) edges
           j: = j+1;
           If E_MST U{e_ij} is not a cycle. //An edge which do not form cycle when
           added
                 E_MST:≡ E_MST U{e_ij};
                 counter: =counter +1;
     RETURN E_MST;
}
```

Example 4.3: Find the MST using Kruskal's algorithm for given graph (figure 4.8).

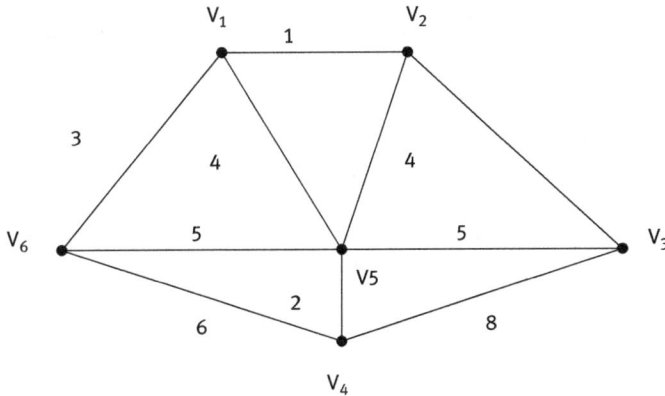

Figure 4.8: Graph.

The minimal spanning tree is given in figure 4.9.

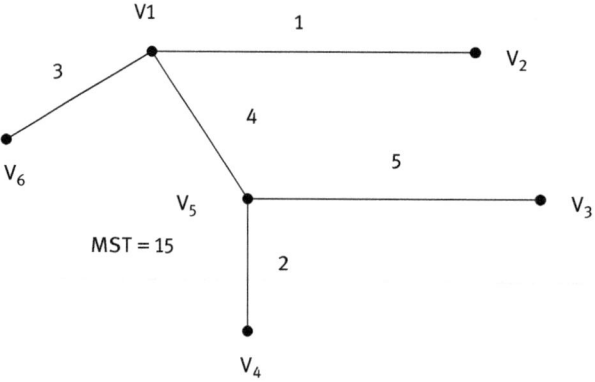

Figure 4.9: MST using Kruskal's algorithm.

Complexity of Kruskal's Algorithm: Complexity of the algorithm is basically depends on the sorting of edge weights of a given graph. Number of edges are $|E|$, so the complexity with an efficient sorting algorithm will be $O(|E|\log_2|E|)$.

4.3.2 Prim's algorithm

The Prim's algorithm to find MST is based on greedy approach. The input to Prim's algorithm is an edge-weighted connected graph (the digraph case is not relevant here) on vertices which we suppose are numbered from 1 to n. At stage r we shall have chosen a set S of r vertices and a spanning tree T on S, all of whose edges lie in G. Initially $S = \{1\}$ and T has no edges. For the rth stage we suppose inductively that S of size $r-1$ and a spanning tree T on S have been found. We select vertices $i \in S$ and $j \notin S$ so that the weight a_{ij} of the edge e from i to j is as small as possible, and adjoin j to S and e to T. When stage n is complete, all vertices have been included, and we have MST for G. MST for example 4.3 is given in figure 4.10 solved using Prim's a algorithm.

Algorithm: *Prim(G)*
// Input: A connected weighted graph
//Output: E_{MST}: The set of edges creating a MST.

```
    {
        V_T : ≡ {v₀};
        E_MST ≡ Φ ;
        For k: = 1 to |V|−1 do          // |V|= n
        Find minimum weight edge e*= (v*,u*), which does not form cycle when include
        in E_MST, among all the edges (v, u) s.t v is in V_T and u is in (v−V_T)
```

$V_T: \equiv V_T U \{u*\}$
$E_{MST}: \equiv E_{MST} U\{e*\}$
Return E_{MST};
}

Complexity of Prim: If adjacency list is used for graph representation and the priority queue is implemented as a min heap, the complexity will be $O(|E|\log |V|)$.

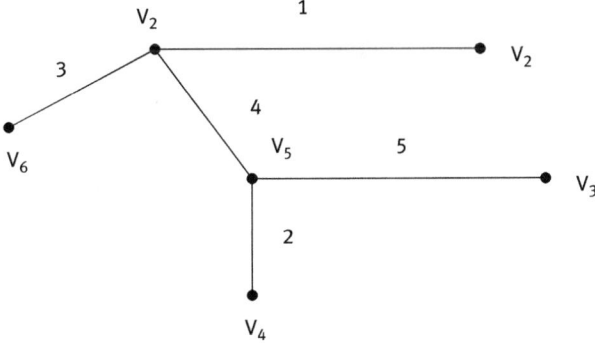

Figure 4.10: MST of Example 4.3 using Prim's Algorithm.

4.4 Shortest path algorithm

This section includes algorithms to find shortest paths between all pair of vertices and from one source to all other vertices in a graph.

4.4.1 Warhsall's algorithm

Warshall's algorithm finds matrices $M_0, M_1, M_2 \ldots \ldots M_n$, where for each in the (r,s) th entry $M_m(r,s)$ of M_m is the least length of a path from vertex r to vertex s where all intermediate vertices if any lie in $\{$ 1,2, ...,m$\}$, again M_n is the desired matrix and this time we may find the entries of M_m for m > 0 by recursive equation

$$M_m(r, s) = \min\{M_{m-1}(r, s), M_{m-1}(r, m) + M_{m-1}(m, s)\}$$

Algorithm: *Warshall ()*

```
{
    For p = 1 to |V| do
    M[p, p]: = 0;
    for p : 1 to |V| do
    {
        for q: = 1 to |V| do
        {
            for r: 1 to |V| do
            {
                M[q,r]: = Min(M[q,r], M[q,p] + M[p,r]);
            }
        }
    }
}
```

Example 4.4:

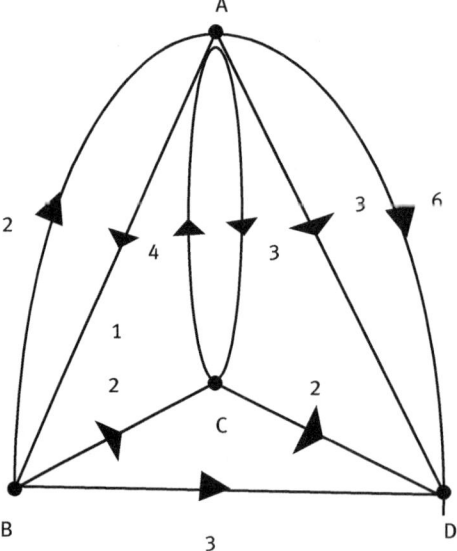

Figure 4.11: Graph.

The matrix will be

$$M_0 = \begin{array}{c|cccc} & A & B & C & D \\ \hline A & 0 & 1 & 3 & 3 \\ B & 2 & 0 & 2 & \infty \\ C & 4 & \infty & 0 & 2 \\ D & \infty & 3 & \infty & 0 \end{array}$$

$$M_1 = \begin{bmatrix} 0 & 1 & 3 & 3 \\ 2 & 0 & 2 & 5 \\ 4 & 5 & 0 & 2 \\ \infty & 3 & \infty & 0 \end{bmatrix} \longrightarrow \{1\} \longrightarrow \text{Set of intermediate vertices}$$

$$M_2 = \begin{bmatrix} 0 & 1 & 3 & 3 \\ 2 & 0 & 2 & 5 \\ 4 & 5 & 0 & 2 \\ 5 & 3 & 5 & 0 \end{bmatrix} \longrightarrow \{1,2\} \longrightarrow \text{Set of intermediate vertices}$$

$$M_3 = \begin{bmatrix} 0 & 1 & 3 & 3 \\ 2 & 0 & 2 & 5 \\ 4 & 5 & 0 & 2 \\ 5 & 3 & 5 & 0 \end{bmatrix}$$

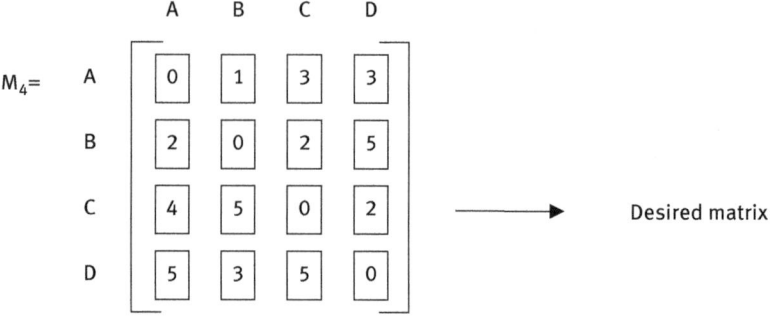

4.4.2 Floyd Warshall's algorithm

In Floyd Warshall's algorithm, a matrix called distance matrix L_{nxn} is used to store the length of shortest path between vertices, where pq^{th} cell denotes the shortest path length between vertex v_p to v_q and $1 \leq p$, $q \leq n$. One advantage of this algorithm is that it can be applied to both directed and nondirected graph. The matrix L of a weighted graph with n vertices is computed through a sequence of n x n matrices.

$$L^{(0)}, L^{(1)}, L^{(2)} \dots \dots L^{(r)} \dots \dots L^{(n)}$$

Matrix L^r (r = 0, 1, ... n) is equal to the length of the shortest path among all the paths from the pth vertex to qth vertex with each intermediate vertex if any, number not more than r. The time complexity of algorithm is $O(n^3)$ due to three nested for loops.

Algorithm: *Floyd-Warshall(W)*
// Input: The weight matrix W
// Output: The distance matrix L of the shortest path lengths
{
    L: = W // W = $L^{(0)}$
    for r: =1 to |V| do
    {
        for p: =1 to |V| do
        {
            for q: =1 to |V| do
            {
                L[p,q]: =Min (L[p, q], L[p, r]+L[r, q]);
            }
        }
    }
    Return L;
}

Complexity: The running time complexity of Floyd Warshall's algorithm is $O(n^3)$.

Example 4.5:

$$W = \begin{pmatrix} 0 & \infty & 3 & \infty \\ 2 & 0 & \infty & \infty \\ \infty & 7 & 0 & 1 \\ 6 & \infty & \infty & 0 \end{pmatrix}$$

Weight matrix for given graph.
$a_{ij} = 0$ for (i=j)
$a_{ij} = \infty$, if no edge between vertex i and vertex j

$A^{[0]} = W$

$$A^{[1]} = \begin{pmatrix} 0 & \infty & 3 & \infty \\ 2 & 0 & 3 & \infty \\ \infty & 7 & 0 & 1 \\ 6 & \infty & \infty & 0 \end{pmatrix}$$

$$A^{[2]} = \begin{pmatrix} 0 & \infty & 3 & \infty \\ 2 & 0 & 5 & \infty \\ 9 & 7 & 0 & 1 \\ 6 & \infty & 9 & 0 \end{pmatrix}$$

Entry calculation for matrix $a^{[2]}_{3,1}$
$a^{[2]}_{3,1} = Min\{a^{[1]}_{3,1}, a^{[1]}_{3,2}+a^{[1]}_{2,1}\}$
$\qquad = Min\{\infty, 7+2\}$
$\qquad = 9$

Similarly other entries will be computed.

$$A^{[3]} = \begin{pmatrix} 0 & 10 & 3 & 4 \\ 2 & 0 & 5 & 6 \\ 9 & 7 & 0 & 1 \\ 6 & 16 & 9 & 0 \end{pmatrix}$$

$$A^{[4]} = \begin{pmatrix} 0 & 10 & 3 & 4 \\ 2 & 0 & 5 & 6 \\ 7 & 7 & 0 & 1 \\ 6 & 16 & 9 & 0 \end{pmatrix}$$

$a^{[4]}_{3,1} = Min\{a^{[3]}_{3,1}, a^{[3]}_{3,4}+a^{[3]}_{4,1}\}$
$\qquad = Min\{9,1+6\} = 7$

4.4.3 Dijkstra algorithm

It is a single-source shortest path algorithm; we have to identify the shortest path from a source vertex to all the remaining vertices of the graph. Dijkstra's algorithm follows greedy approach in that it always chooses the most obviously attractive next step; This algorithm is applicable to graph with positive weights.

For ease, we suppose that the vertices of graph or digraph are numbered from 1 to n, where 1 is the vertex from which minimum distances to the other vertices are required. There are n main steps, at the rth of which we have a set S of r vertices such that the minimum distances from 1 to members of S are definitely known, and the next step is adjoining one of the remaining vertices to S. After step n, we definitely know the minimum distances from 1 to all other vertices, and the algorithm terminates.

In this algorithm, a set S is used to store vertices for which ultimate sorted path length from the source "A" have been found by now. Next the vertex $u \in V-S$ is selected with the minimum shortest path estimate, adds u to S and relaxes all edges leaving u, this process goes on repetitively until we find shortest path to all the vertices.

Algorithm: *Dijkstra(G, A)*
// Input: A graph G and source vertex A.
// Output: The length d_v of a shortest path from A to v and its penultimate vertex pr_v for every v in V

```
{
    for all v in V do
        dv: == ∞: prv = Null;
    Q=V;
    dA: =0;
    VT=Φ ;
    for i = 0 to n-1 do
        c* := Min-Priority(Q);
        VT = VT U{c*};
        For all c adjacent to c* do
        If dc*+ w (c*,c)<dc
        dc = dc*+w (c*,c)
        prc : = c*          //calculate priority
}
```

The time efficiency of Dijkstra's algorithm depends on the data structure used for implementing priority queue and for representing the graph itself. If graph is represented by adjacency list and the priority queue implements as a min heap. Its complexity will be $O(|E|\log_2 |V|)$.

Example 4.6: Find the single-source shortest path of the following graph (figure 4.11).

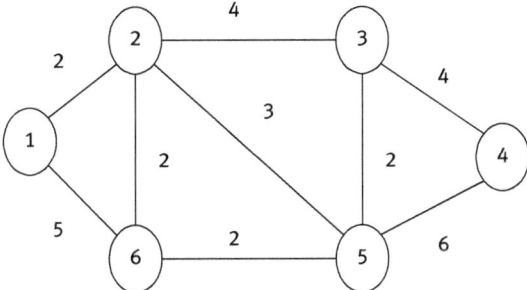

Figure 4.12: Graph G.

S	d_1	d_2	d_3	d_4	d_5	d_6	Array of previous vertices					
							1	2	3	4	5	6
{1}	0	2	∞	∞	∞	5	1					1
{1,2}	0	2	6	∞	5	4	1	2			2	2
{1,2,6}	0	2	6	∞	5	4	1	2			2	2
{1,2,5,6}	0	2	6	11	5	4	1	2	5		2	2
{1,2,3,5,6}	0	2	6	10	5	4	1	2	3		2	2
{1,2,3,4,5,6}	0	2	6	10	5	4	1	2	3		2	2

4.4.4 Bellman–Ford algorithm

The Bellman–Ford Algorithm solves single-source shortest path problem of a given graph that can have negative length weight edges. The function wf is the weight function that returns weight of edge from vertex v_1 to v_2. If there is negative-length cycle, the algorithm signifies that no solution exists. If there is no such cycle, the algorithm produces the shortest paths and their weights.

Example 4.7: Let us take an example, we have graph shown in Figure 4.13. S is the source vertex. We have to find shortest paths from S to all other vertices.

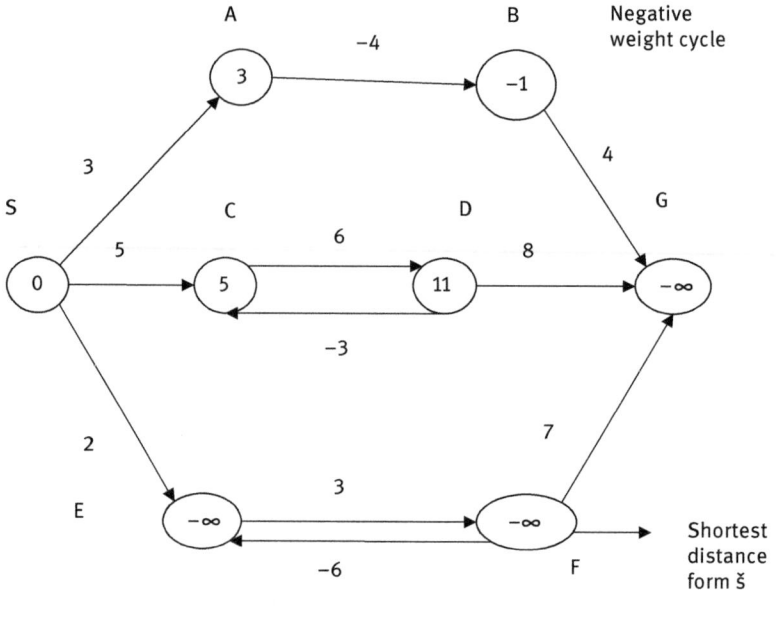

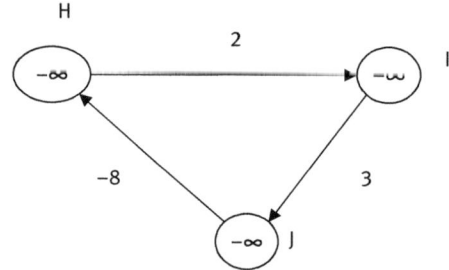

Figure 4.13: Graph.

From S to B, we have only one path, that I s,

$$D(S, B) = wf(S, A) + wf(A, B) = 3 + (-4) = -1$$

To compute path from S to C, we can see that a cycle is formed from S to C:
Length of cycle (C,D,C) = 6 + (−3) = 3 > 0.
So path length will be (S,C) with weight D (S, C) = 5.
Similarly shortest path from S to D is (S,C,D) with weight D(S, D) = 11.

Again from S to E, there is a cycle of length = 3 +(−6) = −3 < 0 (negative weight cycle). So there is no shortest path for S to E. D(S, E) = −∞.
Similarly D(S,F) = −∞.
Because G is reachable from F, and so D(S, G) = −∞
Vertices H, I, and J are not reachable form S, so D(S, H) = D(S,I) = D(S, J) = ∞.

4.4.4.1 Relaxation

Relaxation is a technique that is used in single-source shortest paths algorithms. It is a method that over and over again reduces the maximum value achieved by the actual shortest path weight of each vertex until this maximum value and the shortest path weight are not equal. A variable D(v) is maintained as maximum value of shortest path from source S to v and it is called shortest path estimate. This attribute is maintained by all the vertices of the graph. Shortest path estimates and predecessors are initialized by the following procedure.

Algorithm: *Initialize-Single-Source (G, S)*

```
{
    for all vertices q∈V do
            D[q] := ∞;
                π [q] : = NIL; //Predecessor
    D[S]:= 0;
}
```

The process of relaxing is basically checking whether the shortest path q found so far by going through p can be improved or not, and if so, updating D[q] and π[q].

Algorithm: *Relax (p, q, w)*
```
{
    if (D[q] > D[p]+ w[p,q])
            Then D[q]: = D[p]+ w[p,q];
                π [q]: = p;
}
```

Example 4.8:

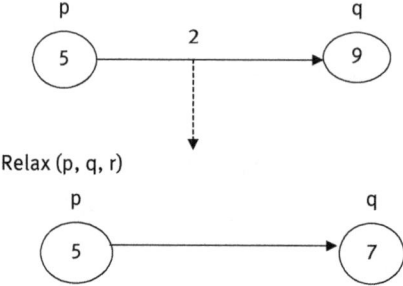

Relax (p, q, r)

D[q] = 9, D[q] + w(p, q) = 7 < D[q]
so new D[q] = 7, π[q]: = p;

Algorithm: *BELLMAN-FORD (G,W,S)*
```
{
    Initialize-Single-Source(G,S)
    for j: = 1 to |V|-1
            do for every edge (p, q)∈E
                    do RELAX (p, q, W);
    for every edge (p, q)∈E
            do if D[q] > D[p] + w (p, q)
                    then return FALSE
    return TRUE                    // no negative weight cycle
}
```

Complexity: The complexity of BELLMAN–FORD Algorithm is O(|V||E|).

Example 4.9: Find the shortest path to all node from node S.

$D[x] = \infty$
$D[S] = 0$
$\pi[v] = NIL$

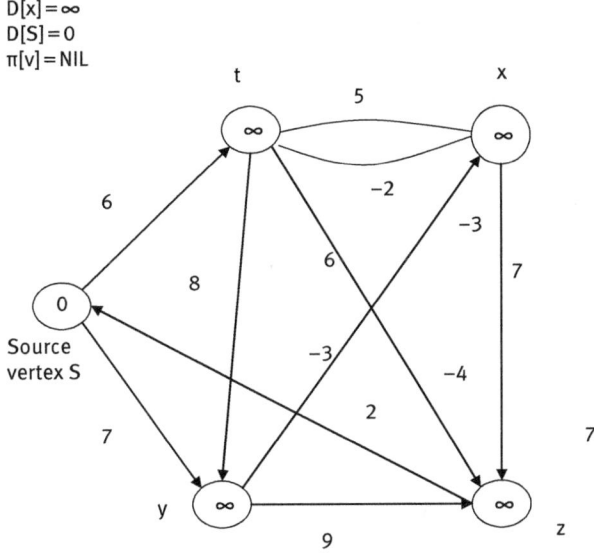

$\pi[t]: = S, \pi[y]: = S, \ D[t]: = 6, \ D[y]: = 7$ (1)

$j = 1$

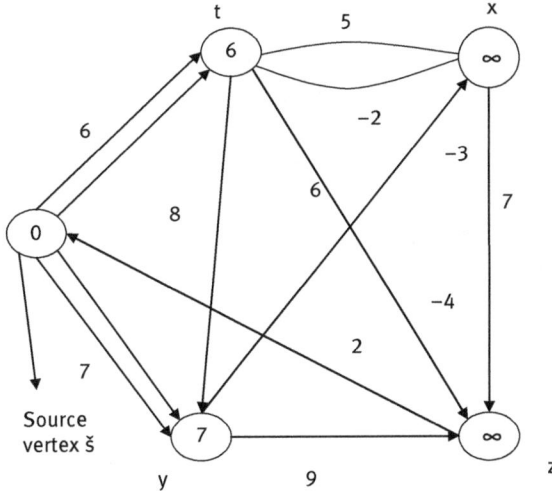

// NUMBER OF ITERATION |V| −1 = 5−1 = 4 FOR i //(2)

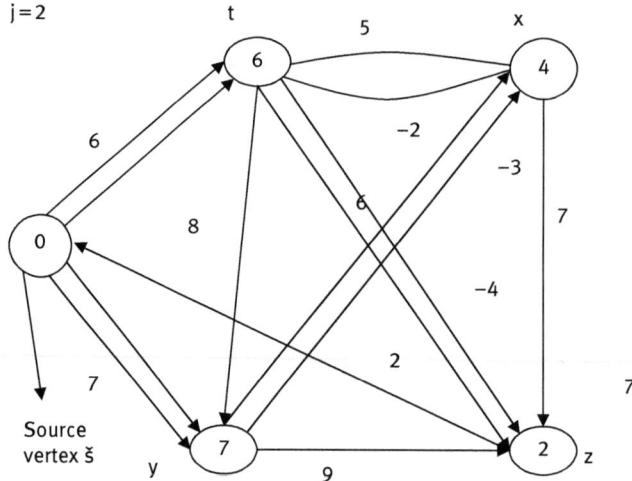

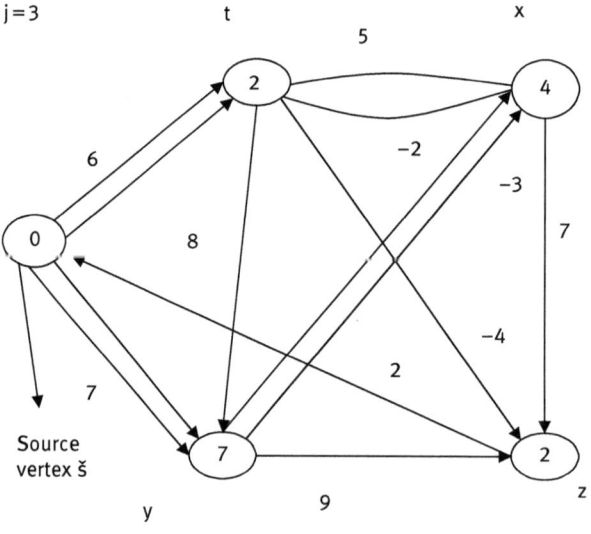

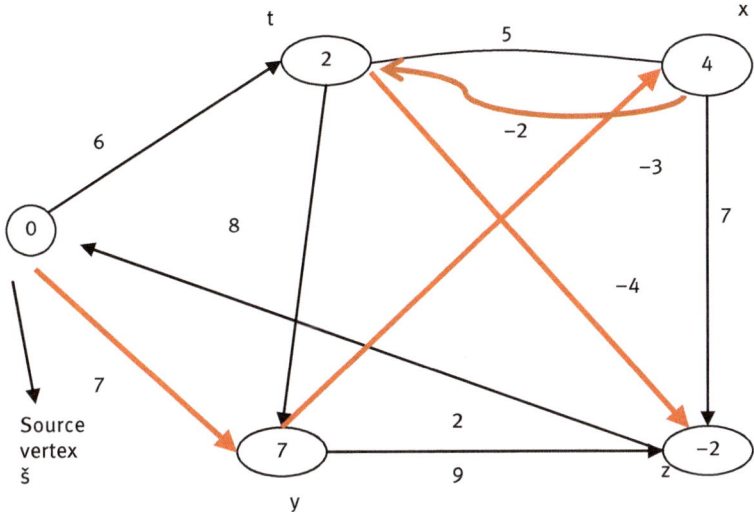

Figure 4.14: Shortest paths from S to remaining vertices.

4.5 Maximum flow

Some real life problems, like flow of liquid through channels, current through cables, and information through communication channels, can be modeled using flow networks. Following are some terminologies used in maximum network flow problems:

Network: It is connected weighted graph in which the edges are the links and every vertex is a station. Weights are assigned to every edge and are known as its capacity.

Capacity: Capacity of an edge is the maximum rate at which the material flows through that edge per unit of time.

Flow Network: A flow network is a directed graph G in which,
- Each edge (p, q) belongs to E, has a nonnegative capacity C(p, q) ≥ 0.
- The two special vertices are source(S) has in-degree 0 and sink (T) has out-degree 0.
- Also there exist a path from source to sink.

In other words, it is a directed graph in which a material flows through the network from source to a sink. Source produces the material and sink consumes the same.

Flow: Let S be the source and T be the sink, then ***flow*** in G is a real valued function F:Vxv → R that satisfies the following three properties (C is the capacity of flow network):

- **Capacity constraint:** The flow in an edge can not exceed the capacity of the edge, that is

$$\text{For all a, b belongs to V we have } F(a, b) < = C(a, b)$$

- **Skew Symmetry:** Flow from vertex a to vertex a is negative of the flow from vertex b to vertex a, that is,

$$\text{For all a, b belongs to V, we have } F(a, b) = -F(b, a)$$

- **Flow conservation (flow in equals flow out):** The rate at which material enters a vertex must be equal to rate at which it leaves the vertex.

Feasible Flow: Total inward flow at intermediate vertex equals to total outward flow at the vertex. A flow that assures the conservation property is called a **Feasible Flow.** Let F' be a feasible flow in a network G. The flow of the network, denoted by F(G) is the sum of flows coming out of the source s.

Let F' be a feasible flow in G. Edge (p, q) is said to be

a) **saturated** if F(p, q) = C(p, q) (see figure 4.15),
b) **free** if F(p, q) = 0,
c) **Positive** if $0 < F(p, q) < C(p, q)$.

4.5.1 Maximum flow and minimum cut

Maximum flow of the network is the feasible flow in a network with capacity such that the value of the flow is as large as possible.

The cut for a flow network G is a set which contains those edges of G, which on deletion stops flow from source S to sink T. The capacity of a cut is equal to the sum of capacities of edges present in the cut.

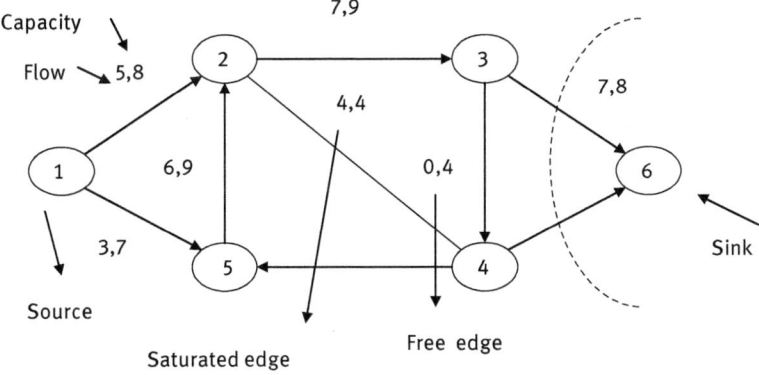

Figure 4.15: Graph.

In max flow problem, we have to calculate the maximum rate at which the material can be flowed from the source to the sink without violating any of the capacity constraints. The three methods to solve max flow problem are as follows:
1) Max flow min cut theorem
2) Ford Fulkerson method
3) Push re-label Method

In a capacitated network, the value of a maximum flow is equal to the capacity of a minimum cut. This is also known as ***max flow min cut theorem***.

4.5.2 Ford Fulkerson method

This is method used for solving the max flow problem. It uses two concepts:
– Residual network
– Augmenting paths

Residual Network: Given a flow network G = (V, E), S is the source node and T is the sink node T. Let f be the flow in G and suppose (p, q) is an edge in G, then the amount of added flow that can be pushed form p to q without exceeding the capacity c of the edge (p, q) is called the residual capacity of edge (see figure 4.16). The residual capacity (rc) of an edge can be defined as follows:
rc(p, q) = c(p, q) – f(p, q) when (p, q) is a forward edge,
and
rc(p, q) = f(p, q) when (p, q) is a backward edge.
Given a flow network G = (V,E) and flow f, the residual network of G induced by f is
G_f = V,E_f) where,

$$E_f = \{(u, v) \in v x v : c_f(u, v) > 0\}$$

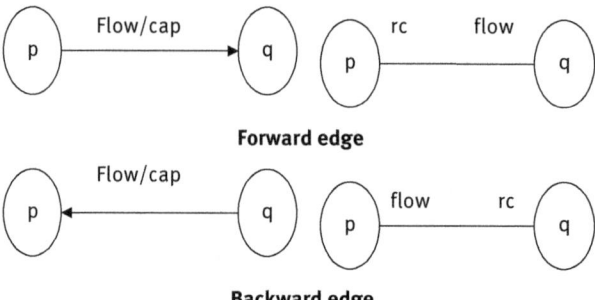

Forward edge

Backward edge

Figure 4.16: Forward and Backward Edge.

Augmenting Path: Given a flow network G and a flow f, an augmenting path P is a simple path from S to T in the residual network G_f, that, every edge (p,q) on an augmenting path admits some additional positive flow from p to q without relating the capacity constraints on the edge.

The residual capacity of augmenting path P is

$$RC_f(p) = \min\{RC_f(p, q) : (p, q) \text{ is in } P\}$$

Algorithm: *FORD-FULKERSON-ALGORITHM (G, S, T)*

```
{
    for each edge (p, q) of graph G
        do f[p, q] := 0
            f [q, p] : =0
    While there exist an augmenting path P from S to T in the residual network Gf
        //The excess flow capacity of an augmenting path equals the minimum of
        the capacities of each edge in the path.
    do RCf(P):= min {RCf (p, q) : (p, q ) is in P}
    for each edge (p, q) in path P
        do f[ p, q ] := f [p, q ] + RCf(P)
            f[ q, p ] := - f [ p, q ] // The while loop repeatedly finds an
            augmenting path P in Gf and augments flow f along P by residual
            capacity RCf(P) when no augmenting path exist, the flow f is maximum
            flow. //
}
```

Example 4.10: Find maximum flow of the graph given in figure 4.17

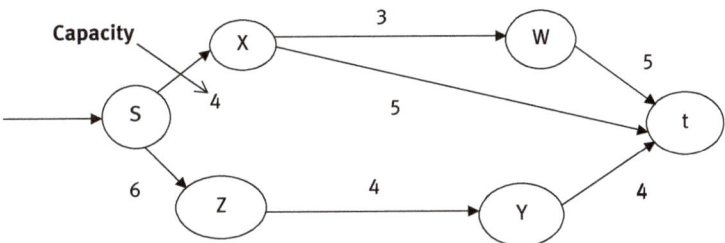

Figure 4.17: Graph

Initially for each edge (p, q) belongs to E set f(p, q) and f(q, p) = 0

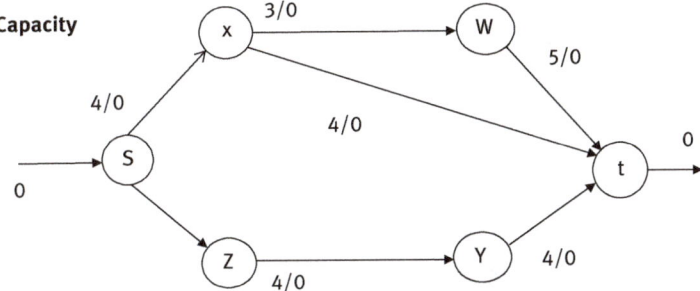

Augmenting Path: s →X → W → t.
Excess capacity of s → X → W →t = min (4,3,5) = 3.
Initially

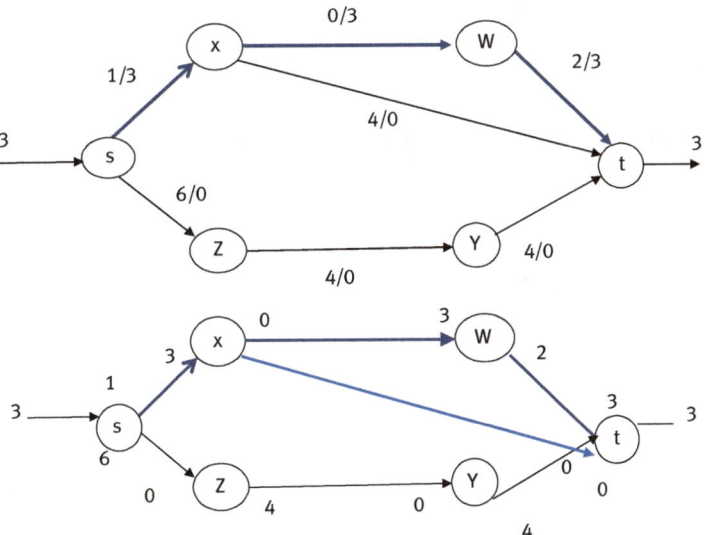

Augmenting path: s –> X –> t
Excess capacity of s –> X –> t = m (1, 5) = 1

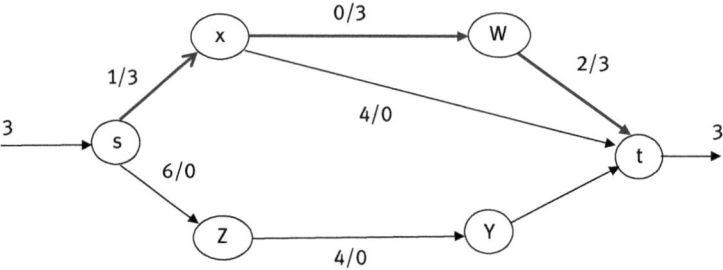

Augmenting path: s –> Z –> Y –> t
Excess capacity of s –> Z –> Y –> t = min (6, 4, 4) = 4

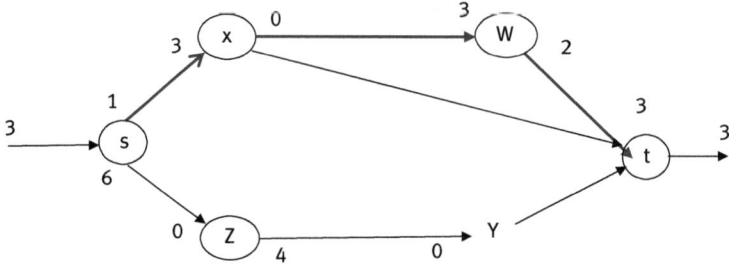

At this point, there are no remaining augmenting paths!
Therefore the flow is maximum = 8. Hence maximum flow of the network = 8

Problem set

1. What do you understand by spanning tree? What is the importance of finding minimal spanning tree?
2. Find MST by applying Kruskal's algorithm to the following graph.

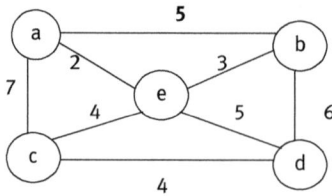

3. Apply Prim's algorithm to find a MST of the following graphs.

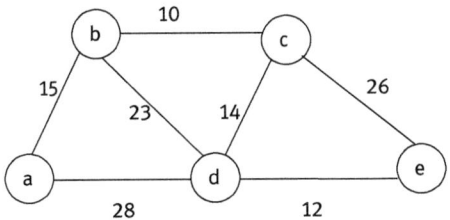

4. Show that prim's algorithm can like Kruskal's algorithm be implemented using heaps.
5. Write pseudo code to determine whether or not a directed graph is singly connected.
6. Use BFS to visit various vertices in the following graph given, taking C as starting vertex.

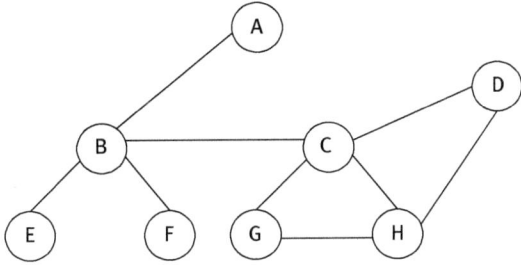

7. Find maximum flow from source 1 to sink 6 in the given flow network.

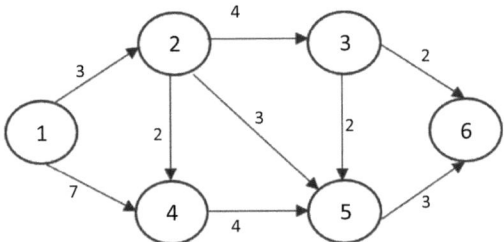

Chapter 5
Number theory, classification of problems, and random algorithms

5.1 Division theorem

Division theorem states that if x and y are integers, $y \neq 0$. Two exceptional integers a and b exist such that

$$x = y.a + b, \text{ where } 0 \leq b < y$$

The quotient is $a = \lfloor x/y \rfloor$. The ultimate result b = x mod y is called the principle remainder of the division. The remainder b is 0 if x is divisible by y.

5.1.1 Common divisor and greatest common divisor

As can be understood by the term common divisor, a number is called a common divisor if it divides two numbers. Let S is the divisor of x and y then it holds the following property:

$$S|x \text{ and } S|y \rightarrow S|(x+y) \text{ and } S|(x-y)$$

In common terms, $S|(xa + yb)$ for integers a and b.

The gcd is the largest common divisor of two integers x and y, not both zero and is denoted by gcd (x, y).

Theorem 5.1: *If there are two integers x and y s.t. x, $y \neq 0$, then gcd of x and y is the smallest positive number of the set {xa + yb: a, b ϵ Z}, where Z is a set of integers.*

Proof: Let us assume that G be an integer that is smallest positive such that

$$G = xa + yb \text{ for some } a, b \in Z.$$

Let q be the quotient and value of q is $\lfloor x/G \rfloor$.

$$x \bmod G = x - q.G$$
$$= x - q(xa + yb)$$
$$= x(1 - qa) + y(-qb)$$

https://doi.org/10.1515/9783110693607-005

Hence *x mod G* is a linear combination of x and y also but as x mod G < G, so we have that x mod G = 0 since G is the least positive such linear combination and so G|x, and similarly G|y. Therefore G is a common divisor of x and y and so

$$\gcd(x, y) \geq G \qquad\qquad\text{(i)}$$

Again gcd(x, y) divides both x and y and G is linear combination of x and y.
   But gcd (x, y)|G and G > 0

$$\Rightarrow \gcd(x, y) \leq G$$

Merging with eq (i) we get

$$\gcd(x, y) = G.$$

5.2 Chinese remainder theorem

Sun-Tzu, a mathematician from China found answer to the problem of identifying a set of integers x that give remainder 2 if divided by 3, give 3 if divided by 5, and remainder 2 if divided by 7. This theorem proves a match between a system of equations modulo a set of pair wise relatively prime moduli and equation modulo their product.

Theorem: Let m_1, m_2, m_3 . . ., m_i , . . ., m_k be relatively prime integers, these are regularly known as *moduli* or *divisors*. Let the product of these numbers up to k^{th} number be M.

According to the theorem, if the m_i are pairwise coprime, and a_1, . . ., a_k are whole numbers, then p is the only integer, such that $0 \leq p < M$ and the remainder of the Euclidean division of *p* by m_i is a_i for every *i*.

The above statements may be formed as follows in term of congruence:

$$p \equiv a_1 (mod\ m_1)$$

$$p \equiv a_2 (mod\ m_2)$$

$$\cdot$$

$$\cdot$$

$$p \equiv a_k (mod\ m_k),$$

and any two such *p* are congruent modulo *M*.

Proof: To establish this theorem, we need to show that a solution exists and it is unique modulo M. First we will construct the solution and then prove that yes it is the solution.

Let $M_{k'} = \frac{M}{m_{k'}}$, $\{M'_k = m_1.m_2.m_3 \ldots m_{k'-1}.m_{k'+1} \ldots \ldots m_k\}$, for $k' = 1, 2, \ldots$ k that is $M'_{k'}$ is the product of all $m_{k'}$ except m_k.

Clearly $(M_{k'}, m_{k'}) = 1$

Also there exists an integer $y'_{k'}$, inverse of $M'_{k'}$ modulo m_k such that

$$M'_k y'_k \equiv 1 \bmod(m_k)$$

To construct the solution for the set of equations,

$$p = a_1 M_1 y_1 + a_2 M_2 y_2 + \ldots \ldots \ldots \ldots + a_k M_k y_k$$

Now we will show that p is the solution of system of equations

Taking p mod$(m_{k'})$, since

$m_J = 0 \bmod(m_{k'})$ when $J \neq k'$

Thus all terms produce remainder 0 when divided by $m_{k'}$ except the term $a_k M'_k y'_k$ So

$$Re\left(\frac{p}{m_{k'}}\right) = \left(\frac{a_k M'_k y'_k}{m_{k'}}\right) + 0$$

Again

$$= Re\left(\frac{a_{k'}(S.m_{k'} + 1)}{m_{k'}}\right) = a_{k'}$$

Which is the solution of k′ th equation.

Since

$$(M_{k'}.y_{k'}) \equiv 1 \bmod(m_k)$$

Thus

$$(M_{k'}.y_{k'} - 1)/m_{k'} = S \,(integer)$$

$$(M_{k'}.y_{k'}) = (S._{mk'} + 1)$$

Similarly, for other equations, thus we have p is the solution to the given system of equations.

To get smallest such p, take p mod (M) and this will be the smallest number which will satisfy the given system of equations.

Solution of Puzzle: Let p be the number of the things

p/3 has reminder 2, Thus $p \equiv 2(\text{mod } 3)$

p/5 has reminder 3, Thus $p \equiv 3(\text{mod } 5)$

p/7 has reminder 2, Thus $p \equiv 2(\text{mod } 7)$

OR

$$\left. \begin{array}{l} p \equiv 2(\text{mod } 3) \\ p \equiv 3(\text{mod } 5) \\ p \equiv 2(\text{mod } 7) \end{array} \right\} \textit{set of } \text{congruent Eq}^n$$

Taking $N = n_1 n_2 n_3 = 3.5.7 = 105$

$$n_1 = N/3 = 35$$
$$n_2 = N/5 = 21$$
$$n_3 = N/7 = 15$$

Again we see that 2 is inverse of 35 (modulo 3), Since

$$35 \times 2 \equiv 1(\text{mod } 3).$$

Also 1 is the inverse of (21 mod 5), so $\because 1 \text{ x } 21 \equiv 1(\text{mod } 5)$

And 1 is inverse of 15(mod 7)

$$\because 1 \text{X} 15 \equiv 1(\text{mod } 7)$$

As per the theorem, $p = a_1 n_1 y_1 + a_2 n_2 y_2 + a_3 n_3 y_3 = 2.35.2 + 3.21.1 + 2.15.1 = 233$

Taking 233 mod(105)= 23

5.3 Matrix operations

5.3.1 Strassen's matrix multiplication

The complexity of product of two matrix of order (nxn) by simple method is $O(n^3)$. Strassen designed a way to compute product of two matrices with only 7 multiplication and 18 addition and subtractions.

// STARSON MATRIX MULTIPLICATION //

X[n : n], Y[n : n]   // each matrix of order n Xn
for a: = 1 to n do
for c : = 1 to n do
Z[a][c]: 0.0;
For b:=1 to n do
Z[a][c]: = Z[a][c] + X[a][b] *Y[b][c];
Return Z;

The complexity of multiplication is simply given by $T(n) = cn^3$ (complexity multiplication).

5.3.2 Divide and conquer for matrix multiplication

We have to perform multiplication of two matrices of dimension x*x. Let us assume that x is a power of 2, that is, there exists a positive integer k s. t. $x = 2^k$

In case x is not power of two then enough rows and column of zero's can be added to both the matrices, so that resulting dimension are of power of 2.

Let us assume that X and Y be each divided into four submatrices, each sub matrix has dimension $x/2 \times x/2$

$$\begin{bmatrix} X_{11}X_{12} \\ X_{21}X_{22} \end{bmatrix} \begin{bmatrix} Y_{11}Y_{12} \\ Y_{21}Y_{22} \end{bmatrix} \quad \{8/sub/matrices\}$$

$$\begin{bmatrix} Z_{11}Z_{12} \\ Z_{21}Z_{22} \end{bmatrix}$$

Then
$$Z_{11} = X_{11}{}^*Y_{11} + X_{12}{}^*Y_{21}$$
$$Z_{12} = X_{11}{}^*Y_{12} + X_{12}{}^*Y_{22}$$
$$Z_{21} = X_{21}{}^*Y_{11} + X_{22}{}^*Y_{21}$$
$$Z_{22} = X_{21}{}^*Y_{12} + X_{22}{}^*Y_{22}$$

If x > 2, matrix Z's elements can be computed by multiplication and addition operations applied to matrices of size $x/2 \times x/2$. To perform matrix multiplication of two matrices 8 multiplications and 4 additions of $x/2 \times x/2$ matrices are performed.

Note: The complexity of addition of two matrices is given by $O(x^2)$.

Since two matrices of order $x/2 \times x/2$ can be added in time cx^2, where c is a constant, and then earlier statement can be denoted by a relation T(x), which is

$$T(x) = \begin{bmatrix} b, x \le 2 \\ 8T(x/2) + cx^2, x > 2 \end{bmatrix}$$

where, b and c are constant.

On simplification we get

$$T(x) \approx O(x^3)$$

These is no improvement over conventional multiplication method.

5.3.3 Strassen's multiplication

Stranssen's designed a way to compute product matrix Z using only of 7 multiplications and 18 additions and subtractions. This algorithm computes seven $x/2 \times x/2$ matrices $Z_1, Z_2, Z_3, Z_4, Z_5, Z_6,$ and Z_7. These matrices can be calculated by performing 7 multiplications and 10 matrix additions or subtractions. Values of these seven matrices are as follows:

$$Z_1 = (X_{11} + X_{22})^*(Y_{11} + Y_{22})$$

$$Z_2 = (X_{21} + X_{22})^*(Y_{11})$$

$$Z_3 = X_{11}{}^*(Y_{12} - Y_{22})$$

$$Z_4 = X_{22}{}^*(Y_{21} - Y_{11})$$

$$Z_5 = (X_{11} + X_{12})^*Y_{22}$$

$$Z_6 = (X_{21} + X_{11})^*(Y_{11} + Y_{12})$$

$$Z_7 = (X_{12} - X_{22})^*(Y_{21} + Y_{22})$$

And

$$Z_{11} = Z_1 + Z_4 - Z_5 + Z_7$$

$$Z_{12} = Z_3 + Z_5$$

$$Z_{21} = Z_2 + Z_4$$

$$Z_{22} = Z_1 + Z_3 - Z_2 + Z_6$$

So the recurrence relation is

$$T(x) = \begin{bmatrix} b, x \le 2 \\ 7T(x/2) + ax^2, x > 2 \end{bmatrix}$$

$$T(x) = ax^2 \left[1 + 7/4 + (7/4)^2 + \ldots\ldots\ldots\ldots(7/4)^{k-1} \right] + 7^k T(1)$$

$$= ax^2 \left(\frac{7}{4} \right)^{log_2 x} - ax^2 + 7^{log_2 x}.b$$

$$= a7^{log_2 x} - ax^2 + x^{log_2 7}.b = O(x^{log_2 7})$$

$$T(x) \approx O(x^{2.81})$$

So, a small improvement has been observed.

5.4 Pattern matching

Let Text[1:p] is a sequence of p elements and Pattern[1:q] is the pattern array where $q \le p$. The elements of *Pattern and Text* are characters drawn from the set of alphabet Σ. Set Σ^* is the set of all finite length strings formed using character z from Σ. The pattern *Pattern* exists with sift z in text *Text* if $0 \le z \le (p-q)$ and Text[z + 1, z + q] = Pattern[1:q], that is Text[z + j] = Pattern[j] for $1 \le j \le q$. If pattern *Pattern* exists with sift z in array *Text*, then z is called a valid shift otherwise invalid. String matching problem is to find all valid shifts.

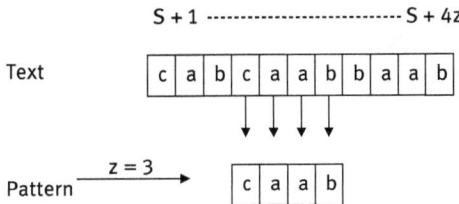

The pattern occurs only once at shift z = 3 in the text.

5.4.1 The Naive algorithm

The worst case complexity of Naive algorithm is O(p−q + 1) m). This bound is too tight (e.g., consider a text a^p and a pattern a^q). If the pattern is half the length of the text (q = p/2), then the worst-case running time is Ω (p^2). Because information gained about the text while trying to match the pattern with shift z is not utilized when one tries to match the pattern with shift z + 1, the naive algorithm is inefficient. Naive algorithm identifies all valid shifts using a loop that ensures the condition Pattern[1:q] = Text[z + 1: z + q] for all (p−q + 1) possible values of z.

NAIVE-STRING–MATCHING–ALGORITHM (Text, Pattern)

1. p: = length[Text]
2. q: = length[Pattern]
3. for z: = 0 to (p-q)
4.   if ({Pattern[1-q] = Text[z+1. . . .,z+q])
5.   then Print " Pattern occurs with shift" z.
This procedure takes time O(p-q+1)q.

5.4.2 The Rabin–Karp algorithm

The Rabin–Karp algorithm is a Monte-Carlo randomized algorithm for matching two strings based on hashing. The idea is to compare hash values rather than strings. Assume we have a hash function h on strings of length m, and let $h_* = h(x)$ be hash of the pattern and $h_s = h(Y_{s+1} \ldots . Y_{s+m})$ the hash of the text at shift z. Check that $h_* = h_s$ for all shift z.

Considerations for h:

- Hash values can be compared quickly
- Fast update: h_{s+1} can be computed quickly from h_s
- Probability of false matches ($h(u) = h(v)$, but $u \neq v$) is bounded

// THE HASH FUNCTION //

For simplicity, consider strings over alphabet $\Sigma = \{0,1\}$. Given a prime p, the hash of a bit string x is obtained by interpreting x as a binary number and dividing by p

$$h(x) = \sum_{i=1}^{m} x_i.2^{m-i} mod\ p$$

Working modulo p allows us to have strings of billions of bits into a few words(e.g., take $p \sim 2^{64}$) and allows us to assume that hash values are compared and updated to constant time. However it introduces the possibility of **false matches:**

– For example, if $p = 17$ then $h(0100101) = h(0010100)$

// UPDATING THE HASH FUNCTION //

If we assume again that $h_s = h(y_{s+1} \ldots . y_{s+m})$, then

$$h_s = \sum_{i=1}^{m} y_{s+i}.2^{m-i}\ mod\ p$$

and

$$h_{s+1} = \sum_{i=1}^{m} y_{s+i+1} \cdot 2^{m-i} \bmod p$$

Thus

$$h_{s+1} = 2\left(h_s - 2^{m-1} \cdot y_{s+1}\right) + y_{s+m+1} \bmod p$$

$$= 2h_s - 2^m \cdot y_{s+1} + y_{s+m+1} \bmod p$$

Now if we recomputed the constant $c = 2^{m-1} \bmod p$, we have a quick way of updating h_{s+1}:

$$h_{s+1} = 2(h_s - c \cdot y_{s+1} \cdot y_{s+m+1}) \bmod p$$

THE RABIN–KARP – MATCHER (Text, Pattern, d, q) // for decimal numbers

```
n : = Length [Text];
m: = Length [Pattern]
h : d^{m-1} mod q                    ( h= 10^{m-1} mod q)
p:= 0;
t_0:= 0;
for i = 1 to m
        do p =(d_p+p[i]), mod q                      {Preprocessing θ(n²)time}
        d_0 := (dt_p +T[i], mod q
        for s: = 0 t_0 (n-m)
                do if p = t_s              { θ (n-m+1)m}
                Then if p[1:-m] = T[s+1: S+m]
                        Then print "pattern occurs with shift's;
                If s < (n-m)
                        Then t_{s+1}:= (d (t_s-T[s+1]h)+T[s+m+1] mod q
```

Example 5.1: Change pattern into hash value. q is the prime number.

P =

2	6	5

d = 10, m = 3, q = 11
for i : = 1 to 3 do
i = 1, p: = (0 + 2) mod 11 = 2
i = 2, p : = (2 × 10 + 6) mod 11 = 4

{ p = 265 mod 11
{ p = 1 mod 11 = 1

i = 3, p: = (4 x 10 + 5) mod 11 = 45 mod 11 = 1

The matching time of the Rabin-Karp algorithm is O(n + m). Because m ≤ n so the matching time will be O(n).

Example 5.2: For working modulo q = 11, search for the pattern p = 26 in the text T = 3141592653589793. Find the number of spurious hits.

Solution: The Given text T[1:16] is

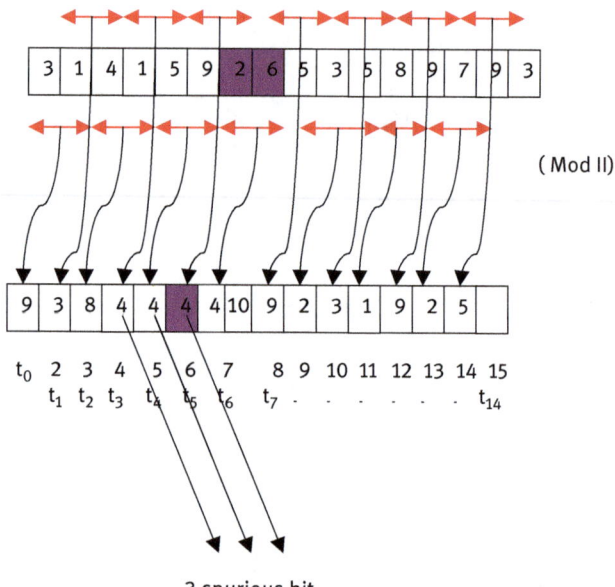

(Mod II)

3 spurious hit

Recursive equation for t_z will be:

$t_{s+1} = d(t_s z - T(s+1)h) + T(s+m+1])\text{mod } q$

$t_1 = (10(31 - - 3 \times 10) + 4)\text{mod } 11$

$= 14 \text{ mod } 11 = 3(\text{mod}11) = 3$

The mod form $t_1 = (10(9-3\text{x}10) + 4)\text{mod } 11$

$(-210+213) \text{ mod } 4$

$= 3 \text{ mod } 4 = 3$

$= -210 + 4 \text{ mod } 11$

$= 14 \text{ mod } 11 = 3 \text{ mod } 11)$

$= 14 \text{ mod } 11 = 3 \ (\text{mod } 11)$

$\{ p = 26$

$p = (26) \text{ mod } 11 = 4$

$t_0 = 31(\text{mod } 11)$

$\quad = 9$

$d = |\Sigma| = 10$

$h = 10^{m-1} \text{ mod } 11$

$\quad = 10 \text{ mod } 11$

$h = 10 \}$

Using algorithm for loop, p=4 and t_0=9.

Now possible values of shift z=0 to 14 (n−m=16−2=14)

Matching the values,

If(p==z)

Then if p[1:2]=T(z+1:z+2)

Then pattern exists otherwise not.

For, $z=0$, $p=4$, $t_0=9$ i.e. $p \neq t_0$
$z=1$, $p=4$, $t_1=5$ i.e. $p \neq t_1$

Similarly for different values of z up to 14, a total of 3 spurious hits encountered and the pattern matches at $z=6$.

5.5 P and NP class problem

We can divide all the decision problems into two classes based on their complexity: class P and class NP. **Class P** problems are the problems that can be cracked in polynomial time by deterministic algorithm. These are also called polynomial. However some problems cannot be cracked in polynomial time, like Hamiltonian circuit traveling salesman problem, knapsack problem, and graph coloring problem. **Class NP** problem are the problems that can be cracked by nondeterministic polynomial algorithm. This class of problem is called nondeterministic polynomial. All class P problems are also class NP problems, so

$$\mathbf{P \subseteq NP}$$

5.5.1 Reducibility

Reducibility is the property of a decision problem. The problem D_1 is said to be polynominally reducible to another decision problem D_2 if there exists a function f that transform instance s of D_1 to instance of D_2 such that
1.   The function f makes all yes instances of D_1 to yes instance of D_2 and all no instance of D_1 into no instance of D_2.
2.   The function f can be calculated by the polynomial time algorithm.

5.5.2 NP complete problem

A decision problem D is said to be NP Complete if
1.   It belongs to class NP.
2.   Every problem in NP class is polynomial reducible to D.

The NP class of problems contains decision problems that are provable in polynomial time. Any problem in P class is also in NP class as already stated.

5.6 Approximation algorithms

In the event that an issue is NP Complete, we are probably not going to locate a poly-nomial time calculation for solving it precisely. Approximation algorithms don't guarantee the best solution. The goal of an approximation algorithm is to come as close as possible to the optimum value in a reasonable amount of time which is at the most polynomial time. An algorithm that returns close to optimal solution is called **approximation algorithm**.

Let i be the instance of some optimization problem and this problem has a large number of possible solutions and the cost of solution founded by approximate algo-rithm is $c(i)$ and the cost of optimal solution is $c^*(i)$. For minimization problem, we are concerned to discover a solution, for which value of the expression $c(i)/c^*(i)$ will be as small as possible. Conversely, for maximization problem, the expression $c^*(i)/c(i)$ value will be as small as possible.

We say that approximation algorithms satisfy following condition:

$$\max(c(i)/c^*(i), \ c^*(i)/c(i)) \leq p(n)$$

Both minimization and maximization problems holds this definition.

Note that the condition $p(n) \geq 1$ always holds.

If the solution generated by approximation algorithm is accurate and optimal then undoubtedly

$$\text{if, } p(n) = 1.$$

5.6.1 Relative error

The relative error of the approximate algorithm can be defined as follows:

$$\text{modulus}[(c(i) - {}^\circ c^*(i))/c^*(i)], \text{ for any input size.}$$

5.6.2 Approximation algorithm for TSP

The nearest neighbor algorithm (greedy approach) is an approximation algo-rithm to solve traveling salesman problem (TSP) and to get optimal solution of TSP, we can apply the dynamic programming approach. Consider the example given in figure 5.1:

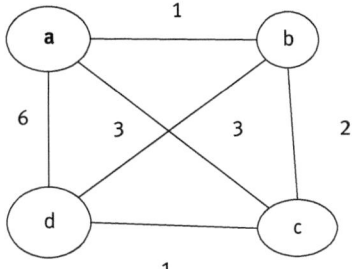

Figure 5.1: A Graph to be solved for TSP.

The nearest neighbor approach, starting from city a,then the nearest neighbor of a is b and nearest neighbor of b is c and then c to d and finally return to vertex a.

$$a\text{-------}^1b\text{------}^2c\text{------}^1d\text{------}^6a$$

The path length of Hamiltonian cycle is 10

But if we apply the DPP, the length of the tour would be 8.

Thus the accuracy ration of this approximation is

$$V(S_a)) = (c(i)/c^*(i)) = 10/8 = 1.25$$

That is, path S_a is 25% larger than the optimal path $S^\star$.

5.6.3 The vertex cover problem

The vertex cover problem comes under the class of NP complete problems. G is an undirected graph with V as the collection of vertices and E as the collection of edges. Vertex cover V' of graph G is a subset of V and if (u, v) is an edge of graph G then either u ∈ V' or v ∈ V' or both. Number of vertices in V is called size of the vertex cover. The problem is to discover a minimum sized vertex cover. This vertex cover will be optimal vertex cover of graph G. It may not be an easy task to discover an optimal vertex cover, but it is not too difficult to discover a vertex cover which is close to optimal.

APPROX-VERTEX- COVER (G)

1. V': = Φ;
2. E' : = E(G)
3. While E' not Φ do
4.   Let take an arbitrary edge(u, v) from the set E'
5.   V':= V' U{u, v}
6.   Remove all the edges that are incident on either u or v from the set E'.
7. return V'.

Example 5.3: Find approximate vertex cover of the graph given in figure 5.2.

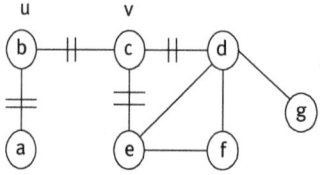

Figure 5.2: Graph.

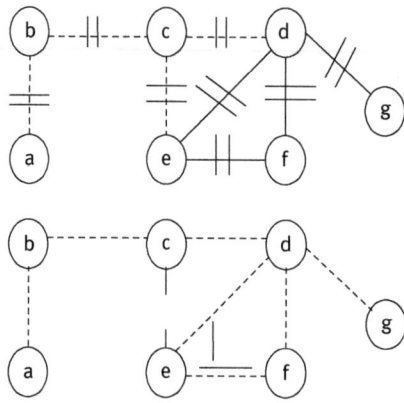

Figure 5.3: Approximate vertex cover of graph given in figure 5.2.

Graph
G ≡ (V,E)
V': = Φ U{a,b}
V': = {a,b}
V': = {a,b} U{e,f}
V': = {a,b,e,f}
V':={ a,b,e,f} U{d,g}
V' = {a,b,e,f,d,g}
E' = Φ{ TERMINATION OF ALGORITHIM}

The optimal vertex cover. The optimal cover is given in figure 5.3.
c*={b, d, e}
size(c*)=3.

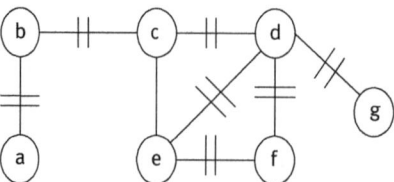

Figure 5.4: The optimal vertex cover.

The time complexity of the above algorithm is O(v + E) using adjacency list to represent E'. This algorithm gives a vertex cover that is at most two times the number of vertices in an optimal cover.

Theorem 5.6.4: APPROX-VERTEX-COVER is a polynomial time 2-approximation algorithm.

Proof: The APPROX-VERTEX-COVER runs ion polynomial time, to prove that APPROX-VERTEX-COVER returns a vertex cover that is at most twice the size of optimal cover.

Let A is the set of edges picked in algorithm APPROX-VERTEX-COVER in order to cover the edges in A, any vertex cover, particular an optimal cover C* must include at least one end point of each edge in A. No two edges in A share an end point since once an edge is picked, all other edges that incident on its end points are deleted from E, thus no two edges in A are covered by any vertex C* of graph G. For C*, we have lower bound

$$|C^*| \geq |A| \qquad\qquad (i)$$

Each execution of line 4 of the algorithm, picks an edge for which neither of its end points are already in C. One edge included in set A mean two vertices included in set C, thus we have

$$C = \{u, v\}$$
$$A = \{(u,v)...\}$$

$$|C| = 2|A| \qquad\qquad (ii)$$

Combining equation (i) and (ii), we have

$$|C| = 2|A| \leq 2|C^*|$$

$$|C|| \leq 2|C^*| \qquad\qquad \textbf{proved}$$

5.7 Deterministic and randomized algorithm

A deterministic algorithm is one that consistently acts a similar way given a similar input; the input totally decides the sequence of calculation execution by the algorithm. Randomized algorithm puts together their conduct with respect to the input as well as on a few arbitrary decisions (see figure 5.5). The equivalent randomized algorithm given a similar input on numerous occasions may perform diverse calculation in every invocation. This implies, among other change that the running time of a randomized algorithm on a given input is not, at this point fixed however itself a random variable. At the point when we investigate randomized algorithm, we are regularly keen on the most pessimistic scenario: worst case running time.

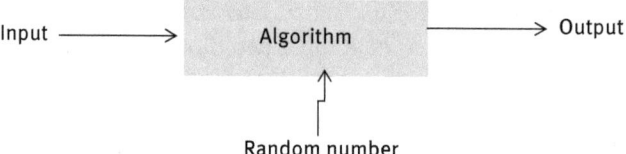

Figure 5.5: Deterministic and randomized algorithm.

5.7.1 The nut and bolt problem

Let us assume that we have n nuts and n bolts of diverse dimensions. Every nut fits with just one bolt and vice versa. The nuts and bolts are all approximately just of the equal size. We wish to discover the nut matches a specific bolt. Then, (n−1) test in worst case whereas in average case, (n/2) test are to be performed.

Let the number of comparisons to discover a match for a sole bolt out of n nuts is denoted by T(n). Obviously when we have only one 1 nut then number of comparisons will be 0 and in case of 2 nuts, the number of comparisons will be one. If number of nuts is greater than 2, then T(n) have value between 1 to (n−1). So expected value of T(n) would be:

$$E\{T(n)\} = \sum_{k=1}^{n-1} k.P_r[T(n) = k].$$

Types of randomized algorithm:
(1) LAS VEGAS: A randomized algorithm that always return a correct result but running time may vary between executions. Ex: randomized quick sort.
(2) MONTE CARLO: A randomized algorithm that terminates in polynomial time but might produce erroneous result. Ex: randomized min cut algorithm.

5.8 Computational geometry

Computational geometry is a part of computer science given to the investigation of algorithm which can be expressed as far as geometry. Some absolutely geometrical issues emerge out of the investigation of computational geometric algorithms, and such problems are additionally viewed as a major aspect of computational geometry. The primary catalyst for the advancement of computational geometry as a control was progress in computer graphics and computer aided design and manufacturing, yet numerous issues in computational geometry are old style in nature and may originate from numerical perception.

Other significant utilizations of computational geometry incorporate robotics, geographic data framework, (geometrical area and search, route arranging), integrated circuit plan (IC geometry structure and confirmation), and computer aided designing.

5.9 The convex hull

The convex hull of a collection Q of points is the least convex polygon P for which each point in Q is either on the boundary of P or in its interior (see figure 5.6). The convex hull of Q is denoted by CH (Q)

$$Q = \{p_0, p_1, p_2, \ldots \ldots \ldots, p_{12}\}$$

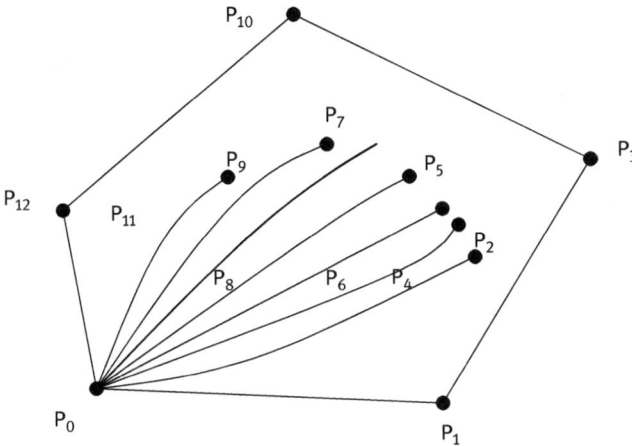

Figure 5.6: The convex hull.

Naturally, we assume that each point in Q is a nail sticking out from a board. The convex hull is then the shape formed by a light rubber band that surrounds all the nails.

GRAHAM-SCAN (Q)

1. Let P_0 be the point in Q with least value of y of coordinate or the left most such point in case of tie.
2. Let $\{P_1, P_2, \ldots \ldots P_m\}$ be the rest of the points in Q sorted by polar angle in counterclockwise fashion around P_0. (if more than one point has a same angle, remove all but the one that is at maximum distance from P_0.)
3. PUSH (P_0, S)
4. PUSH (P_1, S)
5. PUSH (P_2, S)
6. for i = 3 to m
7. do while the angle formed by points NEXT TO TOP (S'),TOP (S'), and P_i makes a nonleft turn.
8. POP(S)
   a. PUSH (P_i, S)
9. Return S

5.9.1 Counterclockwise and clockwise

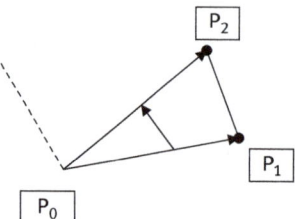

Figure 5.7: Counterclockwise and clockwise Turn.

$$P_0 = (x_0, y_0), P_1 = (x_1, y_1), P_2 = (x_2, y_2)$$

$$P_1 - P_0 = (x_1 - x_0)i + (y_1 - y_0)j$$

$$P_2 - P_0 = (x_2 - x_0)i + (y_2 - y_0)j$$

$$(P_1 - P_0)x(P_2 - P_0) = \{(x_1 - x_0)(y_2 - y_0) - (x_2 - x_0)(y_1 - y_0)\}k$$

If the earlier expression is positive at that point P_0P_1 is clockwise from P_0P_2 and if it is negative then it is counterclockwise.

5.9.2 Finding whether two line segments intersect

A line segment L_1L_2 intersects the other line segment if the end points L_1, L_2 lie on opposite side of the other line. A border line scenario arises when both the lines coincide with each other. The two conditions to check whether two line segments intersect each other are as follows (see figure 5.8):
1. Each line segment straddles the line segment containing the other.
2. An end point of one segment lies on the other line segments.

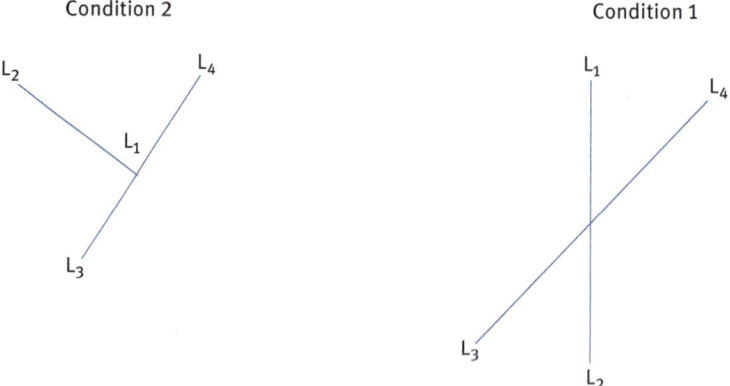

Figure 5.8: Condition for intersecting lines.

To check whether two segments are intersecting or not (see figure 5.8) the algorithm is given below.

SEGMENT INTERSECT (L_1, L_2, L_3, L_4)

1. $c_1 \leftarrow$ direction(L_3, L_4, L_1)
2. $c_2 \leftarrow$ direction(L_3, L_4, L_2)
3. $c_3 \leftarrow$ direction(L_1, L_2, L_3)
4. $c_4 \leftarrow$ direction(L_1, L_2, L_4)
5. if$\{(c_1 > 0)$ and $(c_2 < 0)\}$ or $\{c_1 < 0$ and $c_2 > 0\}$ and $\{c_3 > 0$ and $c_4 < 0\}$ or $\{c_3 < 0$ and $c_4 > 0\}$
6. then return true
7. else if $c_1 = 0$ and on-segment (L_3, L_4, L_1)
8. then return true.
9. else if $c_2 = 0$ and on-segment (L_3, L_4, L_2)
10. then return true
11. else if $c_3 = 0$ and on-segment (L_1, L_2, L_3)
12. then return true
13. else if $c_4 = 0$ and on-segment (L_1, L_2, L_4)
14. then return true
15. else return false

Direction (L_i, L_j, L_k)
1. return$(L_k - L_i) \times (L_j - L_i)$

On-segment (L_i, L_j, L_k)
1. if $\min(x_i, x_j) \leq x_k \leq \max(x_i, x_j)$ and $\min(y_i, y_j) \leq y_k \leq \max(y_i, y_j)$
2. then return true
3. else return false

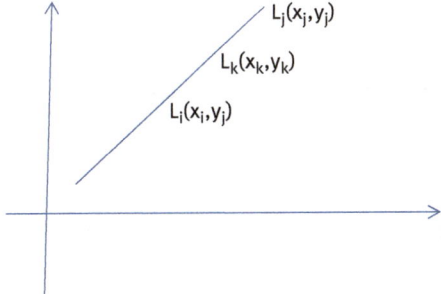

Figure 5.9: Segment intersection.

5.10 Class P problems

An algorithm shows a problem is in polynomial time if its worst-case time efficiency belongs to $O(p(n))$, where $p(n)$ is polynomial of n (input size). These are easy problems since

$$log_2n < n, \; for \; n$$

$$So, \; t(n) \in O(log_2n)$$

$$Also \; give \; t(n) \in O(n)$$

The problems that can be solved in polynomial time are called tractable. Problem that cannot be solved in polynomial time are called intractable. Example: Sorting and searching problems are class P problems.

5.10.1 NP (nondeterministic Polynomial Time) Problems

Class NP is the decision class problems that can be solved by nondeterministic polynomial algorithm. Any problem for which the answer is either "yes" or "no" (0 or 1) is called a decision problem. A problem is NP if you can quickly (in polynomial time) test whether a solution is correct (without worrying about how many rounds it might be to find a solution). Problems in NP class are relatively easy. Example: **Long simple path in a graph.**

Does there exists a simple path from S to t with at least k edges?

If we are given a path, we can quickly look at it and add up the length S, checking that it really is a path with length at least k, this can be done in linear (polynomial time).

5.10.2 Any problem in P is also in NP

Thus P⊆NP

Searching and sorting problems are P problems, so NP problems as well.

5.10.3 NP Hard Problems

Let L_1 and L_2 are problems, and L_1 is reduces to L_2, if there exists a way to solve L_1 by deterministic polynomial time algorithm using a deterministic algorithm that solves L_2 in polynomial time.

Definition: A problem L is NP Hard if and only if satisfiability is reduced to L (satisfiability α L).

5.10.4 NP Complete

A problem L is NP complete if and only if L is NP Hard and L∈NP. Example: Vertex cover problem, travelling sales person problem, Hamiltonian cycle problem, clique problem, etc.

5.10.5 Halting Problem for Deterministic Algorithms

A halting problem is to determine for an arbitrary deterministic algorithm A with input I ever terminates (or enter in infinite loop). This problem is decidable, and there exists no algorithms of any complexity to solve it, so this problem is not in NP. To show

Satisfiability α Time Halting Problem

Construct a problem A whose input is a propositional formula X, if X has n variables then A tries 2^n possible truth assignments and verifies whether X is satisfiable. If it is, A stops and if not then A enters infinite loop.

Hence A halts on input X if and only if X is satisfiable. IF we had a polynomial time algorithm for halting problems then we could solve the satisfiability problem in polynomial time using A, and X is input to algorithm for halting problems. Hence, halting problem is NP hard that is not an NP.

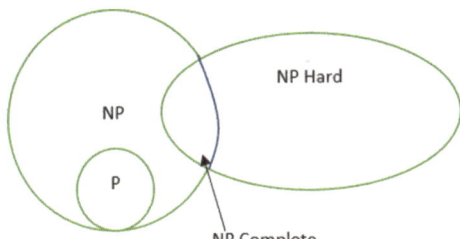

Figure 5.10: Relationship Between P, NP, NP hard and NP complete problems.

5.10.6 The satisfiability problem

Let the Boolean expression be

$$E \equiv a \wedge \neg(b \vee c), E(T) = 1 \text{ i.e. E is true.}$$

E is true if a is true and $\neg(b \vee c)$ is true, when both b and c are false. The expression E is said to be satisfiable if there exist at least one truth assignment T that satisfies E (T) = 1.

$$E \equiv a \wedge \neg(b \vee c)$$

is satisfiable if

$$T(a) = 1, T(b) = 0, T(c) = 0$$

For these values, E(T) = 1.

Theorem 5.10.4: (Cook's Theorem) This theorem states that satisfiability problem is NP complete.

5.11 Clique

A graph G(V, E) is said to be complete graph if all vertices of graph G are adjacent to each other. A clique C of a graph G is a collection of vertices that form complete graph. Size of clique is the number of vertices in C. To discover the largest clique is an NP-hard problem and is known as maximum clique problem. The problem is closely related to vertex cover problem and independent set problem.

Let the set E* be the complement of E (set of edges in G), S is a greatest independent set in the complementary graph G* if S is a maximum clique of G. It follows that V−S is a minimum vertex cover in G*. An example is given in figure 5.11.

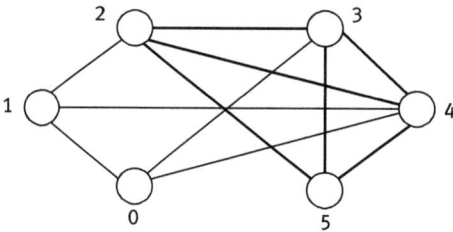

Figure 5.11: Vertex 2-3-4-5 forms a clique of size 4.

5.11.1 Clique Decision Problem

The problem is to find a clique of a particular size (say k) in a given graph G(V, E). It is generally called k-clique problem.

5.11.2 Non-Deterministic Algorithm for Clique Decision Algorithm

This algorithm finds a set of k distinct vertices then check whether these vertices form a complete graph G. The graph G is represented as adjacency matrix and |V|=n.

Algorithm: *DCK(G, n, k)*

S= Φ

i=1 to k do

{

```
    q=choice(1, n)
    if q∈S then failure()
    S=SU{q}
} // for all pairs (p, q) such that p∈S and q∈S and p≠q.
If (p, q) is not an edge of G then failure()
Else success().
```

Theorem: Prove that clique decision problem is NP Complete.

Proof: We will prove this in two steps:

(i) Show that clique problem is in NP.

(ii) 3CNF (3SAT) can be reduced to clique decision problem

Suppose we are given a clique then it is simple to check that whether it is clique of size k, by counting the number of vertices in the clique and it would take polynomial time thus k-clique decision is NP problem.

For second part, we will show that 3CNF (3SAT) can be reduced to clique decision problem.

Consider a Boolean expression,

$$E = C_1 \land C_2 \land C_3 \land \ldots\ldots\ldots\ldots C_r \ldots\ldots \land C_k$$

These are k clauses in E and each clause $C_r = l_1^r \lor l_2^r \lor l_3^r$, where l_1, l_2, l_3 are literals. Then expression

$E(T)=1$, there exists a clique of size k in the corresponding graph G (V, E).

Construction of Graph: For each clause $C_r = l_1^r \lor l_2^r \lor l_3^r$, we place triple V_1^r, V_2^r, V_3^r in V(G) and after this we will get V(G) containing 3k vertices and for E(G) we will connect

(i) V_i^r and V_j^s, (r≠s) {No connection in same triple}

(ii) Also connect all the literals in different triples but not the literals with negation of each other.

Now E = G(V, E)

Since E(T) = 1 and

$E = C_1 \land C_2 \land C_3 \land \ldots\ldots\ldots C_r \ldots\ldots : \land C_k$

and

$C_r = l_1^r . l_2^r . l_3^r$

Thus in each clause C_r, we have at least a literal with the value 1 for E(T)=1.

For each clause pick a vertex corresponding to literal having value 1. And then we have a set of k vertices V'. From the construction of graph G(V, E) all these vertices are connected with each other by one edge and thus V' is a clique of size k.

Again graph G(V, E) has a clique V' of size k by construction of graph G(V, E). We know that no connection in same triple and the literals are not connected with their negations. Since we have a clique of size k, thus we have a vertex in clique corresponding to a clause. Map the literals in each clause for which we have vertex in clique to 1. For each clause, we have one literal which has truth value (i.e. 1) and thus E(T)=1, i.e. E is satisfied. It means clique decision problem is NP Hard, since it is NP too. Therefore it is NP complete problem.

Problem set

1. By taking a condition where all of the elements of a given pattern are different, can we amend the original naive string matching algorithm so that its results will be better for these types of patterns. If we can, then what are the changes to original algorithm?
2. Write down randomized algorithm for quick sort.
3. Rewrite the quick sort algorithm so that it takes a comparison function cmp: $\alpha \times \alpha$ → order and a sequence of type α seq, and only uses the comparison once when comparing the pivot with each key. The type order is the set {Less, Equal, Less}.
4. Solve and explain how 0/1 knapsack problem using approximation algorithm.
5. Explain the satisfiability problem.
6. Explain the convex and concave regions.
7. Write a note on randomized algorithms.
8. Explain the computational geometry.
9. Explain Class P, NP NP hard and NP Complete with help of example.
10. What is clique, explain clique decision problem and max clique problem.
11. Give a nondeterministic algorithm to clique decision problem.
12. State and prove Chinese Remainder Theorem.
13. With working modulo q=13 how many spurious hits does Rabin Karp matcher encountered in text T=4 9 6 2 8 3 4 2 6 5 2 9, when looking for p=6 2.
14. Write a short note on matrix multiplication.
15. Prove that vertex cover problem is NP Complete.

Chapter 6
Tree and heaps

6.1 Red–Black Tree

A Red–Black Tree (RBT) is a kind of binary search tree that holds property of self-balancing. Each node in RBT has color as an extra field; color may be red or black. A RBT (see figure 6.1) must satisfy the following properties:

(i)   The root node always has a black color.

(ii) A nil is considered to be black colored.

(iii) ***The Rule of Black Children:*** The children of red node are black.

(iv) ***The Rule of Black Height:*** For each node in RBT, there is black height which is an integer (**bh(v)**) such that each path from node v to any of its descendant leaves containing same number of black nodes. This is known as the ***black height*** of node v. Black height of root is black height of the tree.

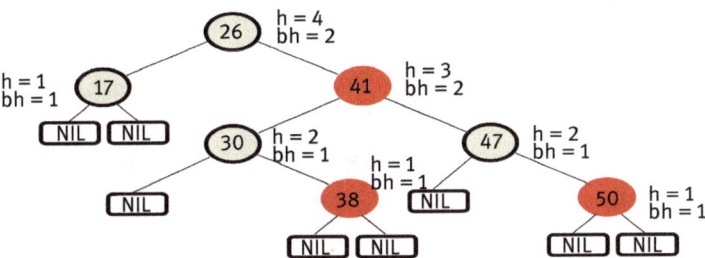

Figure 6.1: Red–Black Tree.

(v) ***Node Height:*** The number of edges in the **longest** route to a descendant leaf node.

(vi) ***A Node's Black Height:*** The number of nodes that are black (including NIL) on the route from the node itself to a leaf node {node itself is not counted}.

Each node of the tree contains fields colour, key, left, right, and a pointer p which indicates, if child or parent of a node doesn't exists then pointer field of the node contains value NIL.

In Red-Black tree, the key bearing nodes are known as internal nodes.

Theorem 6.1.1: A Red-Black tree with n internal nodes has height at most 2 log$_2$ (n+1).

Proof: To prove the theorem, first we show that subtree rooted at any node x contains at least $2^{bh(x)}$-1 internal nodes. We will prove this by mathematical induction on height of node x, if height of node x is 0then node x must be a leaf node and

https://doi.org/10.1515/9783110693607-006

subtree rooted at node x contains at least $2^{bh(x)}-1$ $\{=2^0-1=0\}$ internal nodes. It means, for height 0, statement is valid.

Now consider a node x that has a positive height and is an internal node with two children. Each child has black height either bh(x) or bh(x)–1, depending on its colour is red or black respectively. Since the height of child of node x is less than the height of x itself, we can apply the inductive hypothesis to say that each child has at least $2^{bh(x)-1}-1$ internal nodes. Thus a subtree rooted at x contain at least $(2^{bh(x)-1}-1)+(2^{bh(x)-1}-1)+1=2^{bh(x)}-1$ internal nodes. Thus, the statement is valid.

To prove the theorem, let h be the height of the tree then at least half of the nodes on any path from the root to a leaf node including root must be black. Thus, the black height of the root must be at least h/2. So,

$n \geq 2^{h/2}-1$ (n is number of internal nodes in the tree)

$n+1 \geq 2^{h/2}$

$\log_2(n+1) \geq h/2$

$h \leq 2\log_2(n+1)$ **proved**

6.2 Operations on RBT

This section discusses two operations on RBT, *insertion* and *deletion* of node. Operations left-rotate and right-rotate are used in these operations.

6.2.1 Rotation

Rotation is a confined process in search tree. Let us assume that we have to rotate about the node m and its right child n is not nil. Node n will be the root of the tree. The tree rooted by m will be left child of n and left child of n will be right child of m after performing left rotation (see figure 6.2). The following algorithm given is for left rotation. It is assumed that right[m] ≠ nil, and parent of root node parent is nil.

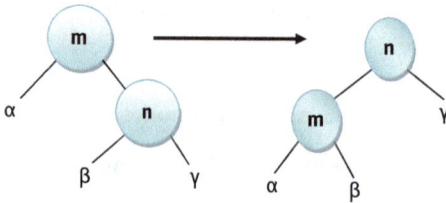

Figure 6.2: Left rotation.

Algorithm: *Left-Rotate()*

```
{
    n ←right[m];
    right[m] ← left[n];
    Parent[left[n]] ←m;
    parent[n]<-parent[m].
    If Parent[m] = nil [Tree]
            Then root [Tree] ← n;
    else if m = left {Parent[m]}
            Then Left Parent [m] ←n;
    else right{Parent[m]} ←n;
    left[n] ←m;
    Parent[m] ←m;
}
```

Time complexity of both the rotations left and right operations is O(1) as only the pointers are changed.

6.2.2 Insertion

Time complexity of inserting a new node in RBT is $O(\log_n)$ time. Algorithm for insertion of a node in tree is given later. Suppose we have to insert node c (for which key [c] = v, left [c] = NIL and right [c] = NIL) in the tree *Tree*. *RB-INSERT-FIXUP(Tree, c)* procedure is used to recolor nodes and perform rotations. A node c is passed as a parameter to the procedure.

Algorithm: *RB-INSERT(Tree, c)*

```
{
  b ← NIL;
  a ← root [Tree];
  while (a is not NIL)
       do b ← a;
       if (key[c] < key[a])
          Then a ← left[a].
       else a ← right [a];
  p [c] ←b;
  if b = Nil ;
          Then root[Tree] ← c;              // Tree is empty
        else if (key[c] < key[b])
          Then left [b] ← c;
    else right[b] ← c;
  left[c] ←nil;   right[c] ←nil;
```

```
        color[c] ←RED;
        RB-INSERT-FIXUP (Tree, c);
}
```

After this, if there is any color variation then the following *RB-COLOR-FIXUP(Tree,c)* given fixes them:

Algorithm: *RB-COLOR- FIXUP (Tree,c)*

```
{
    While color[parent[c]] = RED
        do if parent[c] = left[parent[parent[c]]] // Left of Parent of (Parent of c)
            Then b ← right parent[parent[c]]
                if (color[b] = RED)
                    Then color [parent[c]] ← BLACK
                    color[b] ←BLACK
                    color[parent[parent[c]]] ← RED
                    c ← parent[parent[c]]
                else if c = right[parent[c]]
                    Then c ← Parent[c]
                        Left-Rotate (Tree, c)
                    color[parent[c]] ← BLACK
                    color[parent[parent[c]]] ← RED
                    Right-Rotate (Tree,parent[parent[c]])
                else (Same as then clause with "right" and left" exchanged)
    Color [root[Tree]] ← BLACK
}
```

Example 6.1: Insert the values 42, 39, 32, 13, 20, 9 into an empty RBT.

(1) Insert 42

(2) Insert 39

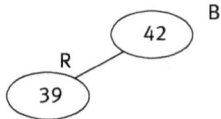

(3) Insert 32

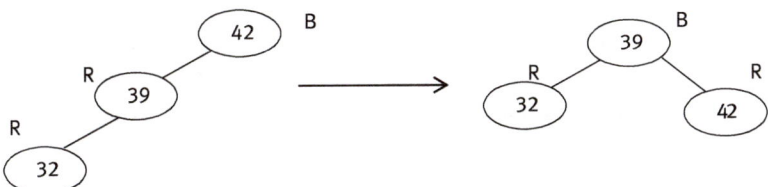

(4) Insert 13

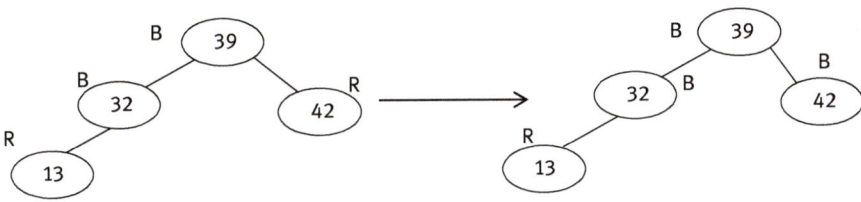

(5) Insert 20

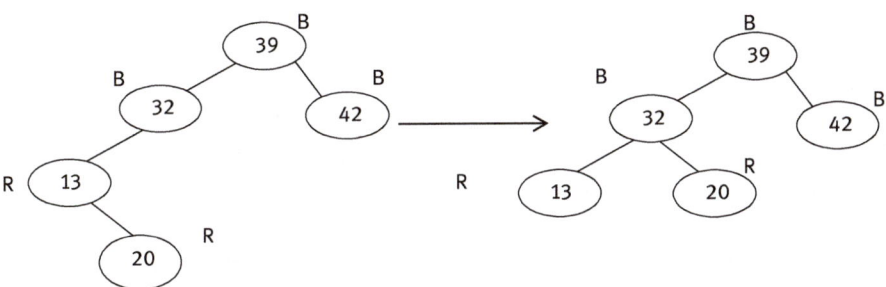

(6) Insert 9

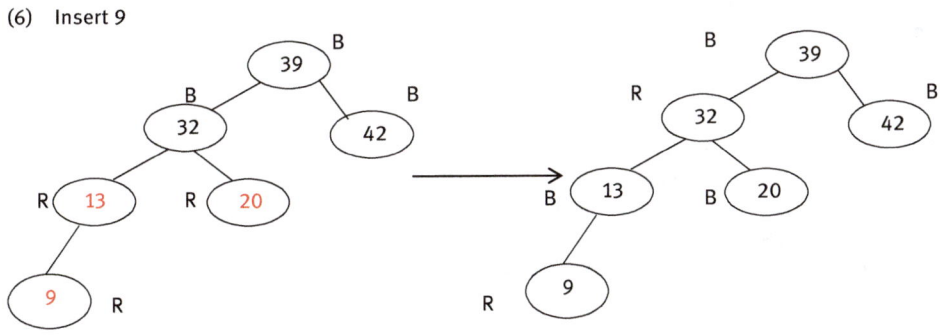

Last one is the final tree.

6.2.3 Deletion

Like the other basic operations on a n-node RB Tree, deletion of a node takes time O(log n).

Algorithm: *RB-DELETE(T, c)*
If left[c]=nil or right[c]=nil then
        b←c;
else
        b←tree-successor[c];
if left[b]≠nil then
        a←left[b];
else
        a←right[b];
parent[a]=parent[b];
if parent[b]=nil then
        root[tree] ←a;
else if b=left[parent[b]] then
        left[parent[y]] ←a;
    else
        right[parent[y]] ←a;
if b≠c then
        key[c] ←key[b]; {copy b data in c}
if color[b]=black then
        RB-DELETE-FIXUP(Tree, a)
Return b;

The call to RB-DELETE-FIXUP(Tree, a) is only if b is black. If b is red the red black property still holds then y is spliced out for the following reasons:
(i)   No black height in the tree has changed.
(ii) No red node have been made adjacent and since b couldn't have been the root if it has red colour, the root remain black.

RB-DELETE_FIXUP(Tree, a)
While a≠root[tree] and color[a]=black
        Do if a=left[Parent[a]] then
                w←right[Parent[a]];
        if color[w]=red then
                color[w] ←black;
                color[Parent[a]] ←red;
                LEFT-ROTATE(Tree, Parent[a]);
                w←right[Parent[a]];
        if color[left[w]]=black and color[right[w]]=black
                color[w] ←red;
                a←Parent[a];
        else if color[right[w]]=black then
                color[left[w]] ←black;

```
            color[w] ←red;
            RIGHT-ROTATE(Tree, w);
            w←right[Parent[a]];
            color[w] ←color[Parent[a]];
            color[P[a]] ←black;
            color[right[w]] ←black;
            LEFT-ROTATE(Tree, P[a]);
            a←root[Tree];
      else (Same as then clause with right and left exchanged)
            color[a] ←black;
```

6.3 B-tree

B-trees are a type of balanced search trees. Concept of B-tree was proposed by Rudolf Bayer and Ed. McCreight. The leaf nodes of B-tree are at the same level. Number of child of internal nodes may vary but within a fixed range. For example, in a 3–4 B-tree, an internal node may have only 3 or 4 children. No frequent rebalancing is required in B-trees because of the flexibility in number of child nodes. Wastage of space is the disadvantage of B-tree since it is not necessary that the nodes must be full. Following are the properties of B-tree:

(i)   Each node a has three fields:

    a)   Number of currently stored keys in node a, that is, $n[a]$.

    b)   These keys are in ascending order, that is,

$$key_1[a] \leq key_2[a] \leq - - \leq key_n[a]$$

    c)   If it is a leaf node then the field Leaf[a] is TRUE otherwise it is FALSE.

(ii) $n[a] + 1$ pointers are also in each node. These pointer $\{c_1[a], c_2[a], \ldots . c_{n[a]+1}[a]\}$ points to child node of a. Since there is no child node for leaf node, c_i fields of leaf nodes are not defined.

(iii) The keys of node a $\{key_i[a]\}$ split the ranges of keys saved in each sub tree of node a. If k_i is any key stored in the sub tree with root $c_i[a]$ then

$$k_i \leq key_1[a] \leq k_2 \leq key_2[a] \leq \ldots \leq key_{n[a]}[a] \leq key_{n[a]+1}$$

(iv) Each leaf node is at same depth like other leaf nodes and this is called height (h) of the tree.

(v) Minimum and maximum number of keys stored in a node is bounded. A fixed integer $t \geq 2$ is used to express these bounds. This is called the ***maximum degree*** of B-tree.

    a)   Lower bound on the keys a node (other than root node) is ***(t–1) keys***. Each nonleaf node except root thus has a minimum of t child nodes. In case of nonempty tree, the root must have no less than one key.

b) Upper bound on the keys a node is **(2 t–1) keys**. So a nonleaf node can have maximum 2 t child nodes. If a node contains exactly (2 t–1) keys then the node is said to be full.

The following figure Figure 6.3 shows the B-tree of height 2. All the nodes contain 1,000 keys.

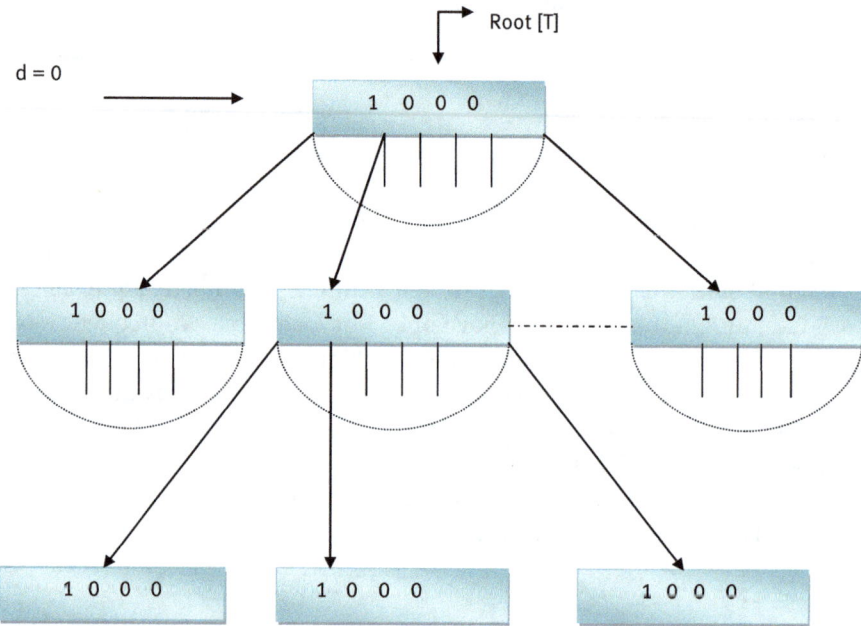

Figure 6.3: B-tree of height 2.

6.3.1 Searching key *k* in B-tree

B-Tree-Search procedure computes the smallest i such that the ith key is more than or equal to a. If the ith key is equal to a, then the search is done, otherwise set z to the ith child.

Algorithm: *B-Tree-Search (z, a)*

```
{
    x = 1;
    while (x ≤ n[z] and a > key_x[z])
        do x = x + 1;
    if (x ≤ n[z] and a = key_x[z])
        then return (z, x);
    if (leaf[z] = TRUE)
```

```
        then return NIL;
    else Disk-Read(cₓ[z]);
        return B-Tree-Serach (cₓ[z], a);
}
```

Example 6.2: Search the key k = 11 in the following B-tree given in figure 6.4.
Solution: $n[x] = 3$, $key_1 = 5$, $key_2 = 9$, $key_3 = 14$, and $k = 11$.

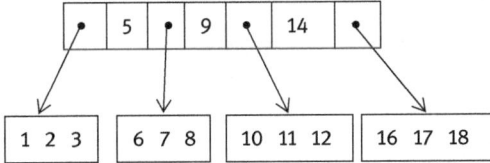

Figure 6.4: B-tree.

For i = 1,
While (i ≤ n[x] and k > keyᵢ[x])
That is, 1 ≤ 3 and 11 > $key_1[x]$ {TRUE}
So i = i + 1, that is, i = 2.
Again, 2 ≤ 3 and 11 > $key_2[x]$ {TRUE}
So, i = i + 1, that is, i = 3.
Now 3 ≤ 3 and 11 > $key_3[x]$ {FALSE}
Since condition for while loop is not correct, checking will be performed at the left of 14. Hence
Disk-Read($c_i[x]$), that is,. Disk-Read($c_3[x]$) and return B-Tree-Search($C_3[x]$, k).
For $C_3[x]$, we have n[x] = 3.
Again for i = 1,
While 1 ≤ 3 and 11 > $key_1[x]$ {TRUE}
So i = i + 1, that is, i = 2.
Again 2 ≤ 3 and 11 > $key_2[x]$ {FALSE}
Check if 3 ≤ 3 and 11 = 11 {TRUE}.
So return (x, i), that is, return (x, 2).

Theorem 6.3.2: If n>=1 then for any n key B tree of hight h and minimum degree t>=2
$$h \leq log_t \left(\frac{(n+1)}{2} \right)$$

Proof: If a B tree has height h, the root contains at least one key and all other nodes contain at least (t–1) keys, thus there are at least 2 nodes at depth 1, at least 2t nodes at depth 2, at least $2t^2$ nodes at depth 3, and so on at depth h there are at least $2t^{h-1}$ nodes.

$$\text{Then } n \geq 1 + (t - 1) \sum_{i=1}^{h} 2t^{i-1}$$

$$= 1 + 2(t^h - 1)$$

$$n \geq 2t^h - 1$$

$$n + 1 \geq 2t^h$$

$$\frac{n+1}{2} \geq t^h$$

$$h \leq \log_t \left(\frac{n+1}{2}\right)$$

6.4 Binomial heap

It is a data structure similar to binary heap but it also supports merge operation of two heaps. Binomial heap is a set of binomial trees. A binomial tree is a recursively defined ordered tree:

(i)   The binomial tree of order 0 has only one node.
(ii) Binomial tree of order greater than 1 (i.e., $k \geq 1$), the binomial tree of order k (B_k) is formed by linking two binomial trees of order k–1 (B_{k-1}). The root of one tree is the leftmost child of the root of the other tree (see figure 6.5).

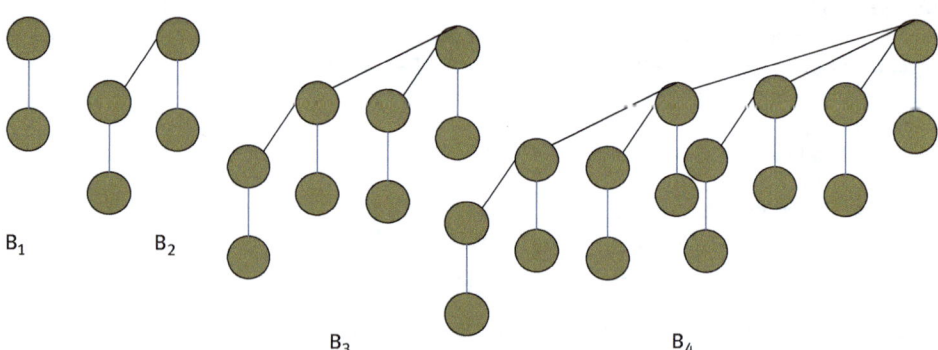

B_1 B_2 B_3 B_4

Figure 6.5: Binomial tree.

Following are the features of binomial tree:

(i)   There are 2^k nodes in the tree and k is height of tree.
(ii) There are just ^kC_i nodes at depth i, where i = 0, 1, 2 . . . k.
(iii) Degree of the root node is k which is more than degree of any other node; besides this the children of the root are binomial tree with order k–1, k–2, ...0 from left to right.

Proof: We will proof it by induction method on k.

For k = 0, each property holds for binomial tree of order 0.

Inductive Step: Let us assume that it holds for k−1 order tree.

(i) Two trees of degree k−1 are linked to form tree with degree k and number of nodes in B_{k-1} is 2^{k-1}, so the tree with degree k has $2^{k-1} + 2^{k-1} = 2^k$ nodes.

(ii) The height of B_k is one greater than the height of B_{k-1}. So, (k−1) + 1 = k.

(iii) Let number of nodes at depth i is N (k, i) of tree B_k. Since number of nodes at depth i in B_k is the number of nodes at depth i in B_{k-1} plus number of nodes at depth i−1 in B_{k-1}.

$$N(k, i) = N(k - 1, i) + N(k - 1, i - 1)$$

$$= {}^{k-1}C_i + {}^{k-1}C_{i-1} = {}^kC_i$$

(iv) Root of B_k is the only node that has greater degree than B_{k-1}. It has one extra child than in B_{k-1}. Degree of root in B_{k-1} is degree k−1 that of B_k is k. By inductive hypothesis, the children of the root of B_{k-1} are roots of $B_{k-2}, B_{k-3}, \ldots B_0$ is linked to B_{k-1}, therefore, the children of the resulting root are roots of $B_{k-1}, B_{k-2}, \ldots B_0$.

6.4.1 Binomial heap

A Binomial Heap is a sequence of binomial trees that satisfies the following min-heap property:

1. Each tree is minimum heap ordered i.e. the key values of the parent is at least as large as or smaller than its children.
2. For any non-negative order k there exists at least on binomial tree whose root has degree k.

For any n node binomial heap:

1. Minimum key contained in the root B_0, B_1.......B_k.
2. It contains at most upper bound of $\log_2 n + 1$ binomial trees.
3. Height is less than or equal to upper bound of log n (see figure 6.6).

Each node of binomial heap contains:

(i) A **key**
(ii) A **degree** for the number of children
(iii) A pointer **child**, that points to leftmost child
(iv) A pointer **sib**, that points to the right sibling
(v) Pointer **p**, that points to the parent

A binomial heap has a root list. A field *head[H]* is used to access the binomial heap. This pointer points to the initial root of the root list of H. If *head[H] = nil*, it means H has no element.

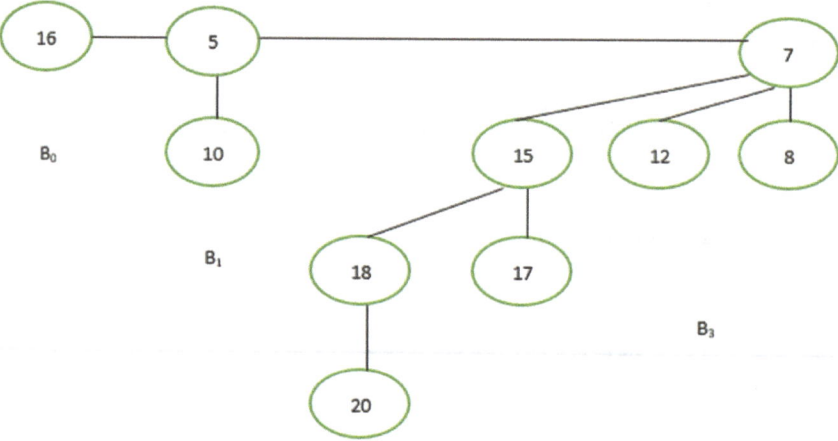

Figure 6.6: Binomial Heap.

6.4.1.1 Creating binomial heap

MAKE-BINOMIAL-HEAP procedure allocate and return an object H, where Head [H] = NIL. The running time is O(1).

6.4.1.2 Finding minimum key

Aim of this procedure is to find the node with minimum key in an n node binomial heap. There is no key with value ∞. Since a binomial heap is a min-heap ordered, a minimum key must reside in the root node. The Binomial-Heap-Min() procedure check all roots which are at most upper bound of (log n)+1, so its running time is O (log$_2$n).

Algorithm: *Binomial-Heap-Min (H)*

```
{
    b←NIL;
    a←head[H];
    min←∞;
    while (a is not NIL)
        do if (key[a]<min)
            then min←key[a];
                b←a;
            a←sib[a];
    return b (min);
}
```

6.4.1.3 Union of two heaps

The procedure for union repeatedly links binomial tree whose roots have same degrees. Let the two trees be rooted at b and c. Node c will become the root of the final tree.

Algorithm: *Binomial-Link (b, c)*

```
{
    p[b]←c;
    sib[b]←child[c];
    child[c]←b;
    degree[c]←degree[c]+1;
}
```

The following procedure unites two heaps T_1 and T_2. Procedure uses *Binomial-Link* and *Binomial-Heap-Merge* as auxiliary procedures. *Binomial-Heap-Merge* merges root list of both the trees into a single and sorted linked list.

Algorithm: *Binomial-Heap-Union (T_1, T_2)*

```
{
    T ← Make-Binomial-Heap ( );
    head [T] ← Binomial-Heap-Merge(T₁, T₂);
    free object T₁ and T₂ but not the list they point to;
    If head [T] = NIL
        Then return T;
    prev->a← NIL;
    a← head [T];
    next->a←sib[a];
    While (next->a is not NIL)
        do if (deg[a] ¹deg[next->a] or
        (sib [next->a] ¹ NIL and deg[sib[next->a]] = degree[a])
            Then prev->a ← a;        // CASE 1 & 2
a←next->a;
        else if (key[a] ≤ key[next->a])
            then sib[a] ← sib[next->a];
            Bionomial-Link (Next->a, a);                // CASE 3
            else if (prev->a = NIL)
            Then head [T] ←next->a;
            else sib [Prev->a] ← next->a;        // Case 4
            Bionomial-Link (a, next->a); // CASE-4
            a← next ->a;
    next->a←sib[a];
    Return T;
}
```

Binomial-Heap-Merge: This procedure merges root list of T_1 and T_2 into a single linked list that is sorted by degree in ascending order.

Algorithm: *Binomial-Heap-Merge(T_1, T_2)*

```
{
    x←head[T₁];
    y←head[T₂];
    head[T₁]← Min-Degree(x,y);
    if head[T₁]=NIL
        return;
    if head[T₁]=y
        then y←x;
    x←head[T₁];
    while (y≠NIL)
        do if sib[x]=NIL
            then sib[x]←y;
                return;
        elseif (degree[sib[x]]<degree[y])
            then x←sib[x];
            else z←sib[y];
            sib[y]←sib[x];
            x←sib[x];
            y←z;
}
```

The time complexity of union operation is O(lg n), where n is the number of nodes in heaps T_1 and T_2.

6.4.1.4 Insertion of a node in binomial heap

Algorithm: *Binomial-Heap-Insert (T, x)*

```
{
    T'←Make-Binomial-Heap();
    P[x]←NIL;
    child[x]←NIL;
    sib[x]←NIL;
    deg[x]←0;
    head[T']←x;
    T←Binomial-Heap-Union(T, T');
}
```

This algorithm runs in O(log n) time. The binomial heap T' is created in O(1) time and union of T' and n node binomial heap in time O(log n).

6.4.1.5 Decrement of a key with a specific value

Following algorithm decreases the key of a node a to a new key d. It displays an error message if d is greater than a's current key.

Algorithm: *BH-Decrease-Key (T, a, d)*

```
{
    If d > key[a]
        Then display message 'decrement not possible';
    key[a]←d;
    b←a;
    c←p[b];
    while c ≠ NIL and key[b] < key[c]
            do exchange key[b]↔key[c];
    b←c;
    c←p[b];
}
```

This procedure takes $O(\log_2 n)$ time, since maximum depth of a is $[\log n]$ { $n = 2^k$, $k = \log_2 n$}.

6.4.1.6 Deletion

This procedure deletes x's key from binomial heap T, we assume that no key has value -∞.

Binomial-Heap-Delete(T, x)
BH-Decrease-Key(T, x,-∞) .
Binomial-Heap-Min(T).

The running time of this procedure is $O(\log n)$.

6.5 Fibonacci heap

Fibonacci heaps are an arrangement of min heap ordered trees. It is not necessary that the trees must be binomial tree. In Fibonacci heap, trees are rooted but unordered. Each node x contains a pointer parent[x] to its parents and a pointer child[x], points to any one of its children. Circular doubly linked list is used to link together the children of x. This list is called *child list of x*.

The two advantages of using circular doubly linked list are as follows:
(i) A node can be removed from this list in O(1) time.
(ii) Two circular doubly linked lists can be concatenated into one list in O(1) time.

A pointer min[H[is used to access the heap H. This pointer points to the root of the tree with minimum key; this is the smallest node of the heap (see figure 6.7). The heap is empty if this pointer is NIL.

6.5.1 Operation on Fibonacci heap

6.5.1.1 New Fibonacci heap: MAKE_FIB_HEAP

This procedure allocates and return Fibonacci heap object H, where n[H]=0, min [H]=NIL. The time requirement of this procedure is O(1).

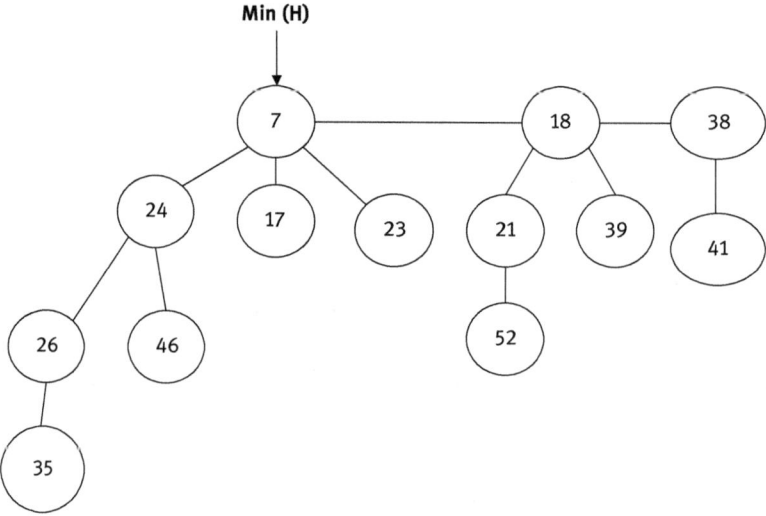

Figure 6.7: A Fibonacci heap.

6.5.1.2 Inserting a node

Insert a node a into Fibonacci heap H, let us assume that the node has already been allocated and that key[a] has already been filled in.

Algorithm: *Fibo-Heap-Insert (H, a)*

```
{
    deg[a]←0;
    parent[a]←NIL;
    child[a]←NIL;
    left[a]←a;
    right[a]←a;
    mark[a]←FALSE;
    concatenate the root list containing a with root list H
```

```
    if min[H] = NIL or key[a] < key[min[H]]
        then min[H]←a;
    n[H]←n[H] + 1
}
```

6.5.1.3 Finding the minimum node

The pointer min[H] points to root of the minimum node of heap, which is also minimum node of heap. It can be directly accessed in O(1) time.

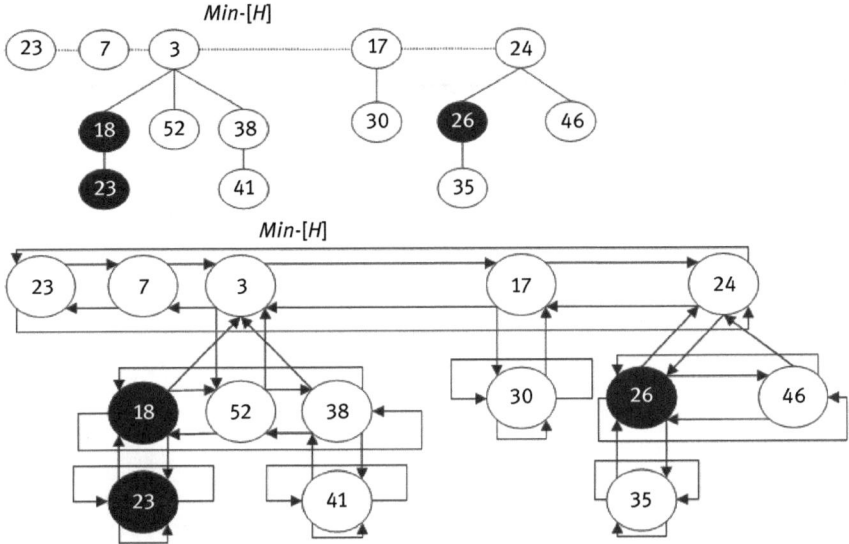

Figure 6.8: Finding the minimum node.

6.5.1.4 Decreasing a Key

This operation assigns a new key k to the key of x such that the new key is less than or equal to current key.

Algorithm: DECREASE_KEY(H, x, k)
//This procedure uses CUT and CASCADE-CUT procedures to maintain properties.
If (k>key[x]) then
        Print "New key is greater than current key"
Else
        Key[x] ←k;
        y←Parent[x];
        if(y≠NIL and key[x]<key[y]) then
                CUT(H, x, y);
                CASCADE-CUT(H, y);

If (key[x]<key[min[H]]) then
       Min[H] ←x;
Return;

Algorithm: CUT(H, x, y)
//This procedure cut off the link between x and parent of x i.e. y, making x a root.
Remove x from the child list of y.
deg[y] ←deg[y]−1;
add x to the root list of H.
Parent[x] ←NIL.
Mark[x] ←False.
Return x.

Algorithm: CASCADE-CUT(H, y)
z←Parent[y];
If(z≠NIL) then
       If(mark[y]=False) then
              Mark[y] ←True;
       Else
              CUT(H, y, z);
       CASCADE-CUT(H, z);
Return;

Problem set

1. Differentiate between RBT and AVL tree.
2. What is the maximum height of a RBT with 14 nodes?
3. Explain the areas of application of RBT.
4. Show the resulting tree of inserting a value 40 into the following RBT given:

```
        30
       /    \
     19      48
            /   \
          36     58
                /
              56
```

5. Write pseudo code for B-TREE-DELETE.

6. Discuss the properties of Binomial trees.
7. Compare the Binomial heap and Fibonacci heap; draw the figure of each separately.
8. Explain with help of an example how to insert a node in a binomial heap.
9. Create a Fibonacci Heap for the list: P=<20, 11, 5, 31, 76, 50, 15, 46, 30, 32>

Chapter 7
Lab session

Appendices

The programming examples for students:

//1. Program for Comparison of growth of functions\\

```
#include <graphics.h>
#include <stdlib.h>
#include <stdio.h>
#include <conio.h>
#include <math.h>
#include <dos.h>

int f(int x)
{
    return(x);
}
int f2(int x)
{
    return(x*x);
}
double flog(double x)
{
    If(x==0)
            return (999);
    else
            return (log(x));
}

void main(void)
{
  /* request auto detection */
  int gdriver = DETECT, gmode, errorcode;
  int midx, midy, i,j,k;
  double lg;
  int mx,my,x,y;
  static int l;
  int m;
  int n = 12345;
  char s[25];
```

https://doi.org/10.1515/9783110693607-007

```
/* initialize graphics, local variables*/
initgraph(&gdriver, &gmode, "");
    mx = getmaxx();
    my = getmaxy();

errorcode = graphresult();
if (errorcode != grOk)
{
    printf("Graphics error: %    s\n",  grapherrormsg(errorcode));
    printf("Press any key to halt:");
    getch();
    exit(1);
}
setbkcolor(6);
setcolor(2);
  x=10;
  y=my-10;//470;
  l=0;
//Print Result\\
printf("\n\tEnter the limit : ");
scanf("%d",&m);
    printf("\n\tX\tf(X)\tf(X*X)\tf(log(x))");
  for (i=0; i<=m; i=i+1)
    {
      j=f(i);
      k=f2(i);
      lg=flog(i);
      if(lg==999)
      {
          printf("\n\t%d\t%d\t%d\tINFINITE",i,j,k);
      }
      else
      {
          printf("\n\t%d\t%d\t%d\t%lf",i,j,k,lg);
      }
      delay(400);
    }

printf("\n\n\tPlease Enter any key to show Graph representation. . . . .. . . ...");
getch();
//Graph Representation\\
    clrscr();
    setbkcolor(6);
    setcolor(2);

line(10, my-10, mx-10,my-10);
line(10, 10, 10, my-10);
```

```
x=10;
y=my-10;//470;
l=0;

for (i=0; i<=m; i=i+1)
{
    j=f(i);
    putpixel(x+j,y-j,1 );

    k=f2(i);
      for(;l<=k;l++)
    putpixel(x+i,y-l,4 );

    lg=flog(i);
    if(lg==999)
    {
        outtextxy(530,450, "log0=Infinite");
    }
    else
    {
    putpixel(x+i,y-lg,6 );
    }
delay(20);
}

//to show the final value

itoa(j, s, 10);
outtextxy(x+j+5, y-j-5, "f(x)="); outtextxy(x+j+50, y-j-5, s);

itoa(k, s, 10);
outtextxy(x+i+5, y-k-5, "f(x*x)="); outtextxy(x+j+65, y-k-5, s);

itoa(lg, s, 10);
outtextxy(x+j+5, y-lg-5, "f(logx)=");outtextxy(x+j+70, y-lg-5, s);

    getch();
    clrscr();
if(j>k&&j>lg)
        outtextxy(200, 200, "f(x) functiom has high order groth");
else if(k>j&&k>lg)
            outtextxy(200, 200, "f(x * x) functiom has high order groth");
else
        outtextxy(200, 200, "f(log(x)) functiom has high order groth");
getch();
closegraph();
}
```

//2. Program for sorting an array using bubble sort\\

```c
//BUBBLE SORTING
#include <stdio.h>
#include <conio.h>

void buuble_sort(int *arr,int n);
int main(void)
{

  int *arr,n;
  clrscr();
  printf("\n\t\tBUBBLE SORTING \n");
  printf("\nEnter the Limit : ");
  scanf("%d",&n);

arr=(int *)malloc(n*sizeof(int)); //allocate memory allocation
    for(i=0;i<n;i++)        //input elements
        scanf("%d",&arr[i]);

    bubble_sort(a,n);

    for(i=0;i<n;i++)     //print sorted array\\
    {
        printf("\t%d",arr[i]);
    }
    getch();
      return 0;
}


//function for bubble sort\\

void buuble_sort(int *arr,int n)
{
    int *arr,k,j,temp,n;
    for(i=0;i<n;i++)
    {
        for(j=0;j<n-k-1;j++)
        {
            if(arr[j]>arr[j+1])    //comparision of elements\\
            {
                temp=arr[j];
                arr[j]=arr[j+1];
                arr[j+1]=temp;
            }
        }
    }
}
```

//3. Program for Heap sort \\

```c
#include<stdio.h>
#include<conio.h>
void reheap(int *a,int p,int s);
void heapsort(int a[],int s);
void main()
{
int *a,i,l;
printf("\nEnter the limit :");
scanf("%d",&l);
a=(int *)malloc(l*sizeof(int));
printf("\nEnter the numbers :");
for(i=0;i<l;i++)
{
scanf("%d",(a+i));
}
heapsort(a,l);
for(i=0;i<l;i++)
{
printf("\n%d",*(a+i));
}

getch();
}
void heapsort(int a[],int s)
{
    int i,t;
    for(i=s/2;i>=1;i--)
            reheap(a,i,s);
    for(i=s-1;i>=1;i--)
    {
        t=a[i];
        a[i]=a[0];
        a[0]=t;

        reheap(a,1,i);
    }
}

void reheap(int *a,int p,int s)
{
    int t=a[p-1];
    int c=2*p;
  while(c<=s)
{
    if(c<s && a[c-1]<a[c])
```

```
    {
         c=c+1;
    }
    if(t>a[c-1])
            break;
    a[c/2-1]=a[c-1];
    c=2*c;
}
    a[c/2-1]=t;
}
```

//4. Program for merge sort \\

```
#include<stdio.h>
#include<conio.h>
void mergesort(int a[],int,int);
void main()
{
    int *arr;
    int i,n;
    clrscr();
    printf("\nEnter the Limit ");
    scanf("%d",&n);

    arr=(int *)malloc(n*sizeof(int));
    for(i=0;i<n;i++)
    scanf("%d",(arr+i));

    for(i=0;i<n;i++)
    printf("\n%d",arr[i]);
    printf("\nAfter sorting:\n");
    mergesort(arr,0,n-1);

    for(i=0;i<n;i++)
    printf("\n%d",arr[i]);
    getch();
}
void mergesort(int a[],int low,int high)
{
    int mid,i,j,p,*t;
    if(low>=high)
    {
         return;
    }
    else
  {
            mid=(int)(low+high)/2;
```

```
            mergesort(a,low,mid);
            mergesort(a,mid+1,high);
            i=low;
            j=mid+1;
            p=0;
            t=(int*)malloc((high-low+1)*sizeof(int));
            while((i<=mid)&&(j<=high))
            {
                    if(a[i]<a[j])
                    {
                            t[p]=a[i];
                            p++;
                            i++;
                }
                  else
                  {
                            t[p]=a[j];
                            p++;
                            j++;
                }
            }
            while(i<=mid)
            {
                    t[p]=a[i];
                    p++;
                    i++;
            }
            while(j<=high)
            {
                t[p]=a[j];
                p++;
                j++;
            }
            for(i=low;i<=high;i++)
            {
                a[i]=t[i-low];
            }
        }
    }
```

//5. Program for quick sort \\

```
#include<stdio.h>
#include<conio.h>

void quick(int a[],int l,int h);
void main()
```

```c
{
    int i,*a,n;
    clrscr();
    printf("\nEnter the Limit");
    scanf("%d",&n);
    a=(int *)malloc(n*sizeof(int));
    printf("\nEnter the Numbers ");
    for(i=0;i<n;i++)
        scanf("%d",(a+i));

    for(i=0;i<n;i++)
        printf("\n%d",*(a+i));
    printf("\nAfter \n");

    quick(a,0,n-1);
    for(i=0;i<n;i++)
        printf("\n%d",*(a+i));
    getch();
}
void quick(int a[],int l,int h)
{
    int i,j,t,p;
    if(l<h)
    {
        i=l+1;
        j=h;
        p=a[l];
        while(1)
        {
            while(a[i]<p)
                    i++;
            while(a[j]>p)
                    j--;
            if(i<j)
            {
                t=a[i];
                a[i]=a[j];
                a[j]=t;
                i++;
                j--;
            }
            else
            {
                break;
            }
        }
```

```
        a[l]=a[j];
        a[j]=p;
        quick(a,l,j-1);
        quick(a,j+1,h);
   }
}
```

//6. Program for radix sort \\

```
#define NUM 100
# include<stdio.h>
#include<conio.h>
#include<dos.h>
#include<math.h>
#include<graphics.h>
void radixsort(int a[],int,int);
void main()
{
  int n,a[20],i,max,d,t;
  clrscr();

  printf("enter the number :");
  scanf("%d",&n);
  printf(" ENTER THE DATA -");
  for(i=0;i<n;i++)
    {
      printf("%d. ",i+1);
      scanf("%d",&a[i]);
    }
    //to find max\\
    max=a[0];
    for(i=0;i<n;i++)
    {
      if(a[i]>=max)
              max=a[i];
    }
    d=0;
    t=max;
    while(max!=0)
    {   ++d;
        max=max/10;
    }
    printf("\nMax = %d\tdigit = %d",t,d);
    radixsort(a,n,d);
    getch();
    }
    void radixsort(int a[],int n,int d)
```

```
{
int rear[10],front[10],first,p,q,exp,k,i,y,j;
struct
{
int info;
int next;
}node[NUM];
for(i=0;i<n-1;i++)
{
node[i].info=a[i];
node[i].next=i+1;
}
node[n-1].info=a[n-1];
node[n-1].next=-1;
first=0;

for(k=1;k<=d;k++)        //consider d digit number
{
for(i=0;i<10;i++)
  {
  front[i]=-1;
  rear[i]=-1;
  }

while(first!=-1)
  {
  p=first;
  first=node[first].next;
  y=node[p].info;
  exp=pow(10,k-1);

  j=(y/exp)%10;
  q=rear[j];
  if(q==-1)
    front[j]=p;
  else
    node[q].next=p;
  rear[j]=p;
  }
  for(j=0;j<10&&front[j]==-1;j++);
  first=front[j];
  while(j<=9)
  {
    for(i=j+1;i<10&&front[i]==-1;i++);
    if(i<=9)
    {
      p=i;
      node[rear[j]].next=front[i];
    }
```

```
        j=i;
      }
        node[rear[p]].next=-1;
  }
//copy into original array
for(i=0;i<n;i++)
  {
    a[i]=node[first].info;
    first=node[first].next;
  }
  for(i=0;i<n;i++)
  printf("\n%d . %d",i+1,a[i]);
}
```

//7. PROGRAM FOR CHAIN MATRIX MULTIPLACTION\\

```
#include<stdio.h>
#include<conio.h>

void cmm(int m[10][10],int s[10][10],int p[10],int n);
int ops(int s[10][10],int i,int j);
void display(int m[10][10],int n);

void main()
{
    int m[10][10]={0},s[10][10]={0};
    int p[10]={0},i,n;
    clrscr();

    printf("\nEnter Total number of matrices :");
    scanf("%d",&n);
    printf("\nEnter the dimensions for matrices :");
    for(i=0;i<=n;i++)
            scanf("%d",&p[i]);

    cmm(m,s,p,n);
    printf("\n\nThe cost of optimal solution matrix \n");
    display(m,n);
    printf("\n\nThe value of split the product \n");
    display(s,n);
    printf("\n\nThe Optimal Parenthesization is \n\n");
    ops(s,1,n);
    getch();
}

//claculate minimum number of scalar multiplaction matrix\\
void cmm(int m[10][10],int s[10][10],int p[10],int n)
```

```
{
    int i,j,k,q,l;
    for(i=1;i<=n;i++)
    {
        m[i][i]=0;
    }
    for(l=2;l<=n;l++)
    {
        for(i=1;i<=n-l+1;i++)
        {
            j=i+l-1;
            m[i][j]=999999;
            for(k=i;k<=j-1;k++)
            {
                q=m[i][k]+m[k+1][j]+(p[i-1]*p[k]*p[j]);
                if(q<m[i][j])
                {
                    m[i][j]=q;
                    s[i][j]=k;
                }
            }
        }
    }
    printf("\nthe number of scalar multiplaction = %d",m[1][n]);
}
//To Solve Optimal Parenthesize Scheme\\
int ops(int s[10][10],int i,int j)
{
    lf(l==J)
    {
        printf(" A%d ",i);
        return(0);
    }
    else
    {
        printf("(");
        ops(s,i,s[i][j]);
        ops(s,s[i][j]+1,j);
        printf(")");
    }
    return(0);
}


//To display the matrix\\
void display(int m[10][10],int n)
{
    int i,j;
    for(i=1;i<=n;i++)
    {
```

```
        for(j=1;j<=n;j++)
        {
            printf("\t%d",m[i][j]);
        }
        printf("\n");
    }
}
```

//8. Program for FRACTIONAL KNAPSACK PROBLEM\\

```
#include<stdio.h>
#include<conio.h>
void main()
{
    int n,i,j;
    float v,t,wgt,w,a,b,max=0;
    float m[10][3],x[10];
    clrscr();
    printf("\nEnter the number of Items " );
    scanf("%d",&n);
    printf("\nEnter Weight and Profit ::");
    for(i=0;i<=n-1;i++)
    {
        scanf("%f%f",&m[i][0],&m[i][1]);
    }
    printf("\nEnter the Knapsack capacity ");
    scanf("%f",&wgt);
    clrscr();
    printf("\n\tItem\tWeight\t\tProfit\n");
    for(i=0;i<=n-1;i++)
    {
        printf("\n\t%d\t%f\t%f",i+1,m[i][0],m[i][1]);
    }

    for(i=0;i<n;i++)
    {
        m[i][2]=m[i][1]/m[i][0];
    }
    printf("\n\n\t---------Calculating Per Unit of Weight----------");
    printf("\n\n\tItem\tWi\t\tPi\t\t(Pi/Wi)\n");
    for(i=0;i<=n-1;i++)
    {
        printf("\n\t%d\t%f\t%f\t%f",i+1,m[i][0],m[i][1],m[i][2]);
    }
    printf("\n\n\t----Decreasing order of Profit Per Unit of Weight------");
    for(i=0;i<n;i++)
    {
```

```
        for(j=0;j<n;j++)
        {
            if(m[i][2]>m[j][2])
            {
                    v=m[i][0]; w=m[i][1]; t=m[i][2];
                    m[i][0]=m[j][0]; m[i][1]=m[j][1]; m[i][2]=m[j][2];
                    m[j][0]=v; m[j][1]=w; m[j][2]=t;
            }
        }
    }
    printf("\n\tItem\tWi\t\tPi\t\t(Pi/Wi)\n");
    for(i=0;i<=n-1;i++)
    {
        printf("\n\t%d\t%f\t%f\t%f",i+1,m[i][0],m[i][1],m[i][2]);
    }
    //fractional logic\\
    for(i=0;i<n;i++)
            x[i]=0;
            v=wgt;
    for(i=0;i<n;i++)
    {
        if(m[i][0]>v)
                break;
        x[i]=1;
        v=v-m[i][0];
    }
    if(i<=n)
        x[i]=v/m[i][0];
    printf("\n\n\t----------- Solution Vector                  -----\n\n");
    for(i=0;i<n;i++)
            printf("\t%f",x[i]);

        //Max Profit
        max=0;
    for(i=0;i<n;i++)
    {
        max=max+m[i][1]*x[i];
    }
    printf("\n\n\n\n\tMaximum profit = %f",max);
    getch();
}
```

//9. PROGRAM FOR 0/1 KNAPSACK PROBLEM\\

```
#include<stdio.h>
#include<conio.h>
void main()
```

```
{
    int wgt,n,i,j,w,a,b,max=0;
    int m[10][2],km[10][10];
    clrscr();
    printf("\nEnter the number of Items " );
    scanf("%d",&n);
    printf("\nEnter values of Input matrix");
    for(i=0;i<=n-1;i++)
    {
        scanf("%d%d",&m[i][0],&m[i][1]);
    }
    printf("\nEnter the Knapsack capacity ");
    scanf("%d",&wgt);
    clrscr();
    printf("\nthe input matrix");
    printf("\n\titem\tweight\tvalue\n");
    for(i=0;i<=n-1;i++)
    {
        printf("\n\t%d\t%d\t%d",i+1,m[i][0],m[i][1]);
    }
    for(i=0;i<=n;i++)
    {
        for(j=0;j<=wgt;j++)
        {
            if(i==0 && j>=0)
                km[i][j]=0;
            else if(i>=0 && j==0)
                km[i][j]=0;
            else if((j-m[i-1][0])<0)
                km[i][j]=km[i-1][j];
            else if((j-m[i-1][0])>=0)
            {
                a=km[i-1][j];
                b=m[i-1][1]+km[i-1][(j-m[i-1][0])];
                if(a>b)
                    km[i][j]=a;
                else
                    km[i][j]=b;
            }
        }
    }
    printf("\n\nKnapsack matrix\n");
    for(i=0;i<=n;i++)
    {
        for(j=0;j<=wgt;j++)
        {
            printf("\t%d",km[i][j]);
        }
        printf("\n");
```

```
        }
        i=n;j=wgt;w=0;max=0;
        printf("\noptimum solution\n");
        printf("\n\n\tinput\tweight\tvalue");
        while(wgt-w>0)
        {
            if(km[i][j]!=km[i-1][j])
            {
                printf("\n\t%d\t%d\t%d",i,m[i-1][0],m[i-1][1]);
                max=max+m[i-1][1];
                w=w+m[i-1][0];
                j=wgt-w;
            }
            i--;
        }
        printf("\n\n\tMaximum value in the knapsack : %d",max);
        getch();
}
```

//10. Pattern Matching for numbers\\

```
#include<stdio.h>
#include<conio.h>

void main()
{
    int T[50],P[50],M[50];
    int i,n,m,s,j,q,p,f,x,TN,PN,PM,nf,line;
    clrscr();
    printf("\nEnter the length of T[1:n] ");
    scanf("%d",&n);
    printf("\nEnter the length of P[1:n] ");
    scanf("%d",&m);
    printf("\nEnter the value of q(modulo)");
    scanf("%d",&q);
    printf("\nEnter the values of T[1:n] : ");
    for(i=0;i<n;i++)
            scanf("%d",&T[i]);
    printf("\nEnter the values of P[1:n] : ");
    for(i=0;i<m;i++)
            scanf("%d",&P[i]);
    i=0;
    while(i<=n-2)
```

```
{
    TN=Todec(T,i,(i+m-1));
    M[i]=TN%q;
    i++;
}
PN=Todec(P,0,(m-1));
PM=PN%q;
clrscr();
printf("\n\nPattern Matching Algorithm\n\n");
printf("\nString T[1:n] : ");
for(i=0;i<n;i++)
    printf("%d ",T[i]);
printf("\n\nString P[1:n] : ");
for(i=0;i<m;i++)
    printf("%d ",P[i]);
printf("\nModulo String : ");
for(i=0;i<n-1;i++)
    printf("%d ",M[i]);
nf=1,line=1;
for(i=0;i<=n-2;i++)
{
    f=0;
    if(PM==M[i])
    {
        f=1,x=0;
        for(j=i;j<(i+(m-1));j++)
        {
            if(T[j]!=P[x])
            {
                f=0;
                break;
            }
            x++;
        }
        if(f==1)
        {
            printf("\n\n%d Pattern Matched at a shift of %d",line,i);
            nf=0;
            line++;
        }
    }
}
if(nf==1)
    printf("\n\nPattern not Matched ");
getch();
}
int Todec(int a[],int l,int u)
{
```

```
    int s=0;
    while(l<=u)
    {
        s=s*10+a[l];
        l++;
    }
    return(s);
}
```

//11. Pattern Matching for string\\

```
#include<stdio.h>
#include<conio.h>

void main()
{
    char *T,*P;
    int *M;
    int i,n,m,s,j,q,p,f,x,TN,PN,PM,nf,line;
    clrscr();
    printf("\nEnter the length of T[1:n] ");
    scanf("%d",&n);
    printf("\nEnter the length of P[1:n] ");
    scanf("%d",&m);
    printf("\nEnter the value of q(modulo)");
    scanf("%d",&q);

    T=(char *)malloc(n*sizeof(char));
    P=(char *)malloc(m*sizeof(char));
    M=(int *)malloc(q*sizeof(int));

    printf("\nEnter the values of T[1:n] : ");
    scanf("%s",T);
    fflush(stdin);
    printf("\nEnter the values of P[1:n] : ");
    scanf("%s",P);
    i=0;
    while(i<=n-2)
    {
        TN=Todec(T,i,(i+m-1));
        M[i]=TN%q;
        i++;
    }
    PN=Todec(P,0,(m-1));
    PM=PN%q;

    printf("\n\nPattern Matching Algorithm\n\n");
```

```
        printf("\nString T[1:n] : ");
        printf("%s ",T);
        printf("\n\nString P[1:n] : ");
        printf("%s ",P);
        printf("\nModulo String : ");
        for(i=0;i<n-1;i++)
                printf("%d ",M[i]);
        nf=1,line=1;
        for(i=0;i<=n-2;i++)
        {
            f=0;
            if(PM==M[i])
            {
                f=1,x=0;
                for(j=i;j<(i+(m-1));j++)
                {
                    if(T[j]!=P[x])
                    {
                            f=0;
                            break;
                    }
                    x++;
                }
                if(f==1)
                {
                    printf("\n\n%d Pattern Matched at a shift of %d",line,i);
                    nf=0;
                    line++;
                }
            }
        }
        if(nf==1)
            printf("\n\nPattern not Matched ");
        getch();
}
int Todec(char a[],int l,int u)
{
    int s=0;
    while(l<=u)
    {
        s=s*10+a[l];
        l++;
    }
    return(s);
}
```

//12. Program for TSM problem\\

```c
#include<stdio.h>
#include<conio.h>
#include<alloc.h>

void tsm(int a[10][10],int n);

void main()
{
    int a[10][10],n,i,j,k;
    clrscr();
    printf("\nEnter the total number of city : ");
    scanf("%d",&n);
    printf("\nEnter the distance of cities\n");
    for(i=0;i<n;i++)
    {
        for(j=i;j<n;j++)
        {
            if(i==j)
            {
                a[i][j]=0;
            }
            else
            {
                printf("\nEnter the distance %d --> %d ",i,j);
                scanf("%d",&a[i][j]);
                a[j][i]=a[i][j];
            }
        }
    }
    printf("\nGraph is\n\n");
    for(i=0;i<n;i++)
    {
        for(j=0;j<n;j++)
        {
            printf("\t%d",a[i][j]);
        }
        printf("\n\n");
    }
    tsm(a,n);
    getch();
}
void tsm(int a[10][10],int n)
{
    int *v,c,i,j,k,min,sum=0;
    v=(int *)malloc(n*sizeof(int));
```

```
    for(c=0;c<n;c++)
    {
        v[c]=0;
    }
    v[0]=1;
    min=a[0][1];
    c=0;
    for(i=0;i<n-1;i++)
    {
        printf("\n\t%d",c);
        for(j=0;j<n;j++)
        {
            if((c!=j)&&v[j]!=1)
            {
                if(a[c][j]<=min)
                {
                    min=a[c][j];
                    k=j;
                }
            }
        }
        c=k;
        v[c]=1;
        sum=sum+min;
        printf(" ---> %d = %d\t\tSum = %d",c,min,sum);

        if(c==n-1)
                min=a[c][c-1];
        else
                min=a[c][c+1];
    }
    min=a[c][0];
    sum=sum+min;
    printf(" \n\t%d ---> 0 = %d\t\tSum = %d",c,min,sum);
    printf("\n\nTotal Distance by Salesman : %d",sum);
}
```

//13. Program for Kruskal's Algorithm \\

```
#include<stdio.h>
#include<conio.h>
int parent [10];
void kruskals(int cost[][10],int n);

void main()
{
    int a,b,u,v,i,j,n;
```

```
    int visited[10],min,cost[10][10];
    clrscr();
    //noofedges=1;
    //mincost=0;
    printf("enter the no. of vertix\n");
    scanf("%d",&n);
    printf("enter the adjacency matrix\n");
    for(i=1;i<=n;i++)
    {
        for(j=1;j<=n;j++)
        {
            scanf("%d",&cost[i][j]);
            if(cost[i][j]==0)
                        cost[i][j]=999;
        }
    }
    kruskals(cost,n);
    getch();
}

void kruskals(int cost[][10],int n)
{
    int a,b,u,v,i,j;
    int mincost=0,noofedges=1,min;
    printf("the minimum cost edges are\n");
    while(noofedges<n)
    {
        min=999;
        for(i=1;i<=n;i++)
        {
            for(j=1;j<=n;j++)
            {
                if(cost[i][j]<min)
                {
                    min=cost[i][j];
                    a=u=i;
                    b=v=j;
                }
            }
        }
        while(parent[u])
                u=parent[u];
        while(parent[v])
                v=parent[v];
        if(u!=v)
        {
            noofedges++;
            printf("\nedge( %d --> %d )",a,min);
            mincost+=min;
```

```
                parent[v]=u;
        }
        cost[a][b]=cost[b][a]=999;
    }
    printf("\nminimum cost= %d",mincost);
}
```

//14. Program for Optimal Binary Search Tree\\

```
#include<stdio.h>
#include<conio.h>

void main()
{
    int *P;
    int i,n,PN;
    clrscr();
    printf("\nEnter the limit ");
    scanf("%d",&n);
    P=(int *)malloc(n*sizeof(int));
    printf("\nEnter the keys : ");
    for(i=0;i<n;i++)
            scanf("%d",P[i]);

    PN=opbst(P,n);

    printf("\n\nPN = %d",PN);
    printf("\n\nBST is :");
    for(i=0;i<n-1;i++)
            printf("%d ",P[i]);

    getch();
}
int opbst(int P[],int n)
{
    int s=0,i,j,k,kmin,min,d,sum=0;
    int c[10][10],r[10][10];
    for(i=1;i<n;i++)
    {
        c[i][i-1]=0;
        c[i][i]=P[i];
        r[i][i]=i;
        c[n+1][n]=0;
    }
    for(d=1;d<n-1;d++)
    {
        for(i=1;i<n-d;i++)
```

```
            {
                j=i+d;
                min=9999;
                for(k=i;k<j;k++)
                {
                    if(c[i][k-1]+c[k+1][j]<min)
                    {
                        min=c[i][k-1]+c[k+1][j];
                        kmin=k;
                        r[i][j]=kmin;
                        sum=P[i];
                    }
                }
            }
        }
        for(s=i+1;s<j;s++)
        {
            sum=sum+P[s];
            c[i][j]=min+sum;
        }
        return(c[1][n]);
}
```

Further reading

Aho, A. V., & Hopcroft, J. E. (1974). *The design and analysis of computer algorithms*. Pearson Education India.

Ahuja, R. K., Magnanti, T. L., & Orlin, J. B. (1993). Network flows: theory, algorithms, and applications.

Baase, S. (2009). *Computer algorithms: introduction to design and analysis*. Pearson Education India.

Baase, S. (2009). *Computer algorithms: introduction to design and analysis*. Pearson Education India.

Baudet, G. M. (1978). *The Design and Analysis of Algorithms for Asynchronous Multiprocessors* (No. CMU-CS-78-116). Carnegie-Mellon Univ Pittsburgh PA Dept Of Computer Science.

Bhasin, H., & Gupta, N. (2012). Randomized algorithm approach for solving PCP. *International Journal on Computer Science and Engineering*, 4(1), 106.

Biswas, S. S., Alam, B., & Doja, M. N. (2013). Generalisation of Dijkstra's algorithm for extraction of shortest paths in directed multigraphs. *Journal of Computer Science*, 9(3), 377–382.

Cormen, T. H. (2009). *Introduction to algorithms*. MIT press.

Dave, P. H. (2009). *Design and analysis of algorithms*. Pearson Education India.

EDIIIQN, I. (2007). Introduction to the design & analysis of algorithms.

Fiat, A. (1998). Online Algorithms: The State of the Art (Lecture Notes in Computer Science).

Kingston, J. H., & Kingston, J. H. (1990). *Algorithms and data structures: design, correctness, analysis*. Sydney: Addison-Wesley.

Kiwi, M., & Soto, J. A. (2015). Longest increasing subsequences of randomly chosen multi-row arrays. *Combinatorics, Probability and Computing*, 24(1), 254–293.

Kozen, D. C. (2012). *The design and analysis of algorithms*. Springer Science & Business Media.

Kumar, V., Grama, A., Gupta, A., & Karypis, G. (1994). *Introduction to parallel computing: design and analysis of algorithms* (Vol. 400). Redwood City: Benjamin/Cummings.

Levitin, A., & Mukherjee, S. (2003). *Introduction to the design & analysis of algorithms* (p. 576). Reading, MA: Addison-Wesley.

Mansi, R. (2010). A New Algorithm for Sorting Small Integers. *International Arab Journal of Information Technology*, 7(2).

Panneerselvam, R. (2007). *Design and analysis of Algorithms*. PHI Learning Pvt. Ltd.

Wang, J. (2011). Fast Algorithm of the Traveltime Calculation Based on Binomial Heap Sorts. In *Advanced Materials Research* (Vol. 267, pp. 857–861). Trans Tech Publications.

www.docslide.us

www.edutechlearners.com

www.learningace.com

https://doi.org/10.1515/9783110693607-008

Index

https://doi.org/10.1515/9783110693607-009